From Ancient Roads to Modern Highways: The Science of Concrete Durability

Parkar

Table of contents

Appendix D: Verification of Durability Specification D1

Appendix E: Ethics Signature Form E1

1. Introduction

1.1 Background and context

Roads represent one of the largest public infrastructure investments in most countries and due to growing road networks, the need for durable infrastructure with adequate service life has been recognised. In terms of the history of civil engineering, the need for roads and transport links was envisaged well before Reinforced Concrete (RC) structures were developed. In saying that, one can hence learn a great deal about concrete as a material for construction from: the so-called 'predecessor' of structures which is roads in addition to structures of antiquity.

The SA road network consists of approximately 750 000 km which represents the 10th largest in the world, with a replacement value in the order of R 2 Trillion (i.e. ZAR 2.10^{12}). The South African National Road Agency SOC Limited (SANRAL) is responsible for maintaining the steadily growing national road network of 21 490 km which is expected to reach a long-term goal of 35 000 km (SANRAL, 2016). A current survey reveals there are 10801 bridges and/or major culverts that are maintained on this road network (SANRAL, 2017).

It is well known that the most common and reliable method to assess in-situ concrete strength is by testing concrete cores that are removed from the structure (Smith, 2017). The former statement holds true for compressive strength as well as the durability of hardened concrete. The latter is tested in South Africa by methods described in South African National Standards (SANS) 3001-CO3-1:2015 Civil engineering test methods – Concrete durability index testing – Preparation of test specimens. Core testing of hardened concrete plays an important role in establishing the durability performance in the case of new and existing RC structures. In the case of new construction, cores can be obtained from three different and distinct stages and hence interpreting the variability in line with the specimen source is a key factor, not covered in the SANS test method i.e. the latter two stages as indicated in *Figure 1-1* undergo field curing regimes which contributes to an additional source of variability. However, in the case of existing construction, cores can only be obtained from one stage, through direct assessment of durability which simplifies interpreting the variability.

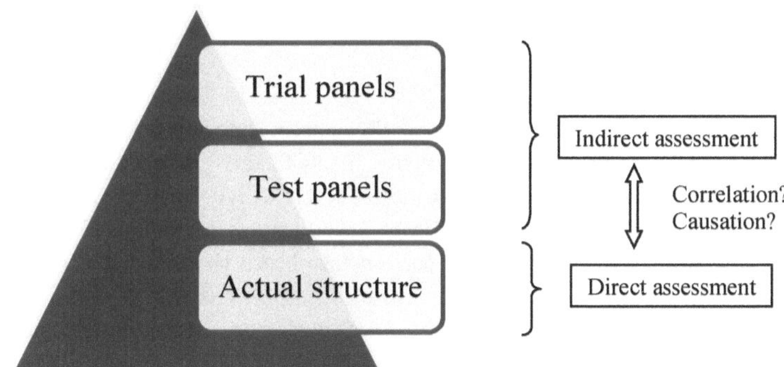

Figure 1-1 Concrete Durability Index (DI) testing (Source: Author)

The independent variables encountered among construction sites are the most important factors that govern concrete durability. As in the Probabilistic methods for durability design : DuraCrete (1999), these three durability variables are defined as:

$$X(cc) = k_e k_c X_0(cc).f(t)$$

with X = material variable

X_0 = reference material variable at time t_0

$f(t)$ = the time dependency factor of the variable

k_e = the environment factor

k_c = the execution factor

Therefore, in RC structures the variability of Durability Index (DI) test results from indirect and direct assessments, is the main concern to ensure structural reliability with sufficient statistical ability. In new construction, cores extracted from the as-built structure are probably more critical than the trial panels cast under laboratory conditions since the former exposes structures to the actual material, manufacturing and testing conditions. However, the importance of laboratory conditions determines the extent from which correlations can be drawn from laboratory results and inferred to actual performance through curing efficiency. Evidently, without the information contained at both these pivotal stages, defining a proper safety margin during a construction period becomes extremely difficult and as such international projects have been directed to this effect.

The testing of core specimens is not complicated; however, the interpretation of the results may be difficult (Smith, 2017). This is another statement that holds true for compressive strength as well as the durability of hardened concrete. The difficulty in interpretation is due to the number of factors that affect concrete durability. Broadly stated, these factors include: concrete composition, execution, environment and the source of specimens. In cases where core testing was performed to assess the in-situ strength of the concrete structures, analysis and interpretation of results were found to be difficult and uncertain (Smith, 2017). Evidently, due to the magnitude of factors affecting concrete durability increases in complexity can be expected.

1.2 Research motivation

Since 2002, SANRAL began to amend the current standard specifications to incorporate additional concrete durability requirements. In 2008, this came with the inclusion of Table 6000/1 : Concrete Durability Specification Targets (Civil Engineering Structures only) in contract documents for performance-based specifications in South Africa for the first time (SANRAL, 2009). Current durability performance-based specifications and quality assurance provisions are contained within Committee of Transport Officials documents (COTO, 2018a; 2018b), respectively, which are currently under revision as working draft chapters.

As such, the performance-based approach in South Africa is a decade into maturity and considerable amounts of actual site data is available and significant for refinements of the DI values in performance-based specifications, the long-term monitoring of RC structures in a full-scale environment along with other research and development (R&D) initiatives.

1.3 Objective and aims

Data modelling can be defined as the process of designing a data model (DM) for data to be stored in a database. Commonly, a DM can be designated into three main types. A conceptual DM defines what the system contains; a logical DM defines how the system should be executed regardless of the Database Management System (DBMS); and a physical DM describes how the system will be executed using a specific DBMS.

The main objective is to design a data model (DM) that is essentially a conceptual and logical representation of the physical database required to ensure durability compliance for RC structures. The advantages of designing this data model (DM) are threefold. It can be used by database developers to create a physical database; it is essential to identify missing and redundant data which lead to errors; and, the Information Technology (IT) infrastructure upgrade as well as maintenance is less expensive and much faster.

The creation of the physical database, through application of the designed data model (DM), will facilitate the current monitoring and management of RC structures due to its ability to deliver project specific numerical summaries of the key parameters influencing concrete durability (i.e. the suite of DI tests, SANS or other), evaluate the results for conformity and acceptance in line with durability specifications by incorporating test data programming or processing and ensure non-conformities are addressed through contractual penalties and/or remedial action. Furthermore, when developed, the database will enable the analysis of laboratory and site-derived DI test results for the implementation of R&D initiatives.

Concrete durability even though less frequently used than compressive strength is arguably the most important concrete design parameter for concrete structures in severe environments, whether new or existing. For the most reliable compressive strength and concrete durability results, the sample must be prepared, tested and the results interpreted strictly according to guidelines stipulated in national standards (Smith, 2017). SANS 3001-CO3-1:2015 provides reliable guidance for the preparation and testing of concrete cores for concrete durability. However, enough guidance for the interpretation and comparison of concrete durability from the three different and distinct specimen sources is not available. Therefore, this study investigated the relationship between site practices (material, manufacturing and testing conditions) and the concrete durability index test results obtained from different sources in the case of new construction. Evidently, this relationship if measured or quantified can be used to draw correlations to actual in-situ performance which has further application in the case of existing structures.

The merit of the database stems from the need to track both experimental and observational data during projects to monitor the variability, assess the data, absolutely and relatively, and hence define a proper safety margin during a construction period. The aims of this study are to:

- Investigate durability design and provisions in codes, standards and specifications from an international context

- Monitor the variability and interplay of observational and experimental conditions on concrete durability properties

- Design a conceptual data model (DM) to establish the basic concepts and scope for the physical database

- Design a logical data model (DM) to add extra information to the conceptual data model (DM) elements

- Identify relations between different input parameters to strengthen the logical data model (DM) elements

1.4 Scope and limitations

The main objective of this study is to design a data model (DM) that is essentially a conceptual and logical representation of the physical database. The data model (DM) which can be used to create the physical database will facilitate organisation and completeness of DI values for site-derived specimens from construction projects located across the country in a systematic manner. The goal of designing such a data model (DM) is to ascertain that the entities or data objects defined are accurately represented.

Alexander, Ballim, & Kiliswa (2013) identified that considerably more work is required to quantify test/sample variability between both batch variability and in-situ variability and that there is a lack of knowledge regarding the magnitude of reduction in values between lab standard cured samples and in-situ achievements. This research is limited to assessing DI values from mainly test panels results, although instances of trial panels and in-situ core results have been reported on, where applicable. A main limitation in past studies was the inability to assess test panel results in relation to what was achieved under standard wet curing conditions in the laboratory which creates difficulty in defining the extremities of construction quality.

In-situ cores are deemed to replicate the conditions found within the actual structure, more so, than test panels. Even though test panels are cast in the same conditions as RC structures which can characterise and convey important information on the trends or correlations to actual in-situ performance, this research stresses the significance of coring the structure when the results obtained from test panels are questionable. Even though this represents a semi-invasive form of testing, sometimes not easily accessible, it is often the only recourse when DI values are unacceptable, which combined with the cover depth achieved can be a reliable indicator for the risk of corrosion. In industry, DI values and the cover depth are the most critical parameters influencing concrete durability and must be captured in the correct places, to make inferences to actual in-situ performance.

However, it is well known that there are still limitations. Khan, Ahmad, & Al-Gahtani (2017) stated that other limitations of the performance-based approach occur mainly due to the following phenomenon:

- Overestimation of exposure class
    - Maturity of internal pore structure decreases penetration ability of chloride ions
    - Decrease in coefficient of chloride diffusion results in less chloride binding
- Overestimation of material resistance
    - Unidirectional chloride diffusion analysis not reflective of actual conditions
    - Threshold chloride concentration not necessarily at point of least cover
    - Rather found at the intersection of all exposed planes (2-D or 3-D effect)
- Exclusion of synergic effect of chloride ingress acting with other failure mechanisms on corrosion initiation

1.5 Book structure

Chapter 1 provides an introduction to the book. A background of the DI approach is discussed as well as progress over the years which mean the performance-based approach in South Africa is now a decade into maturity. The aims of this study are defined in relation to the Durability Index Database (DIDb) followed by setting out the scope and limitations of the study.

Chapter 2 presents a review of literature. Durability design in codes and standards from an international context are discussed, compared and presented in terms of their environmental exposure classes, prescriptive-based methods (design aids and limiting values) and performance-based methods (tests and specifications). A critique of durability provisions is then conducted which examines the different methods for assessing the as-built quality for durability compliance. Following, issues regarding quality control for concrete durability on construction sites are examined such as the stripping of falsework and formwork, cover depth, compaction and curing. A Quality Assurance (QA) scheme is then proposed linked to construction lots which determines test schedules based on project information to assess as-built data in line with performance-based durability specifications such as COTO (2018a; 2018b). The QA scheme is aligned to the South African Road Design Software (SARDS) and existing Pavement Construction Module (PCM).

Chapter 3 focuses on the design of the conceptual data model (DM) which establishes the basic concepts and the scope for the physical database. This process will establish the entities or data objects (distinct groups), their attributes (properties of distinct groups) and their relationship (dependency of association between groups). Database design principles were applied to the main objectives of this study to create a 6-modular structure for the physical database. Database preconditions and requirements were then defined for the Bridge Construction Module (BCM) relating to the data (input, output, user interface, exchange and update), general use, maintenance and extensions.

Lastly, spotlight was placed on the operations and maintenance sector by conducting a database review linking to the conceptual design characteristics of observational and experimental databases presenting some existing database design solutions from international literature and proposing a way forward in the South African context. The first steps in developing the Bridge Construction Module (BCM) was also concluded such as selecting the test methods, setting the sampling frequency, test data programming required and laboratory equipment used, in which the conceptual data model (DM) was signed off.

Chapter 4 focuses on the design of the logical data model (DM) which adds extra information to the conceptual data model (DM) elements by defining the database tables or basic information required for the physical database. This process will establish the structure of the data elements, sets relationships between them and provides foundation to form the base for the physical database. In this chapter, a prototype is presented of the designed data model (DM) founded on 53 basic information database tables. The breakdown of database tables for the six modules is split according to references (1), concrete composition (13), execution (4), environment (7), specimens (2) and test results (26).

Chapter 5 focuses on identifying correlations between different input parameters which adds extra information to the logical data model (DM) elements. Therefore, the relations between the topics defined are strengthened which ultimately determines the extraction of information or output parameters from the physical database according to specification limits. Five different projects which served as input for a total of 1054 Durability Index (DI) test results (4216 test determinations) were used to conduct parametric studies on the most influencing variables affecting concrete durability in Reinforced Concrete (RC) structures.

Chapter 6 provides conclusions that can be derived from the results. Additionally, it suggests the main findings from the parametric studies, key questions that were addressed in the research, as well as further recommendations for practice and future research work.

2. Literature review

2.1 Introduction

There are two fundamental and elementary approaches to durability design. Broadly stated, these are: the prescriptive method; and the performance or model-based method. Li, Zhou, & Chen (2008), describes both methods as being rather complementary in a complete design procedure rather than opposite in nature. In saying that, durability design is hence an iterative process. Hybrid approaches involving both methods allow designers to optimise material constituents in terms of content and composition to formulate a mix design that satisfies the performance criterion. The selection of environmental actions can be somewhat similar for both approaches and is therefore explained in the following sections. The extent and accuracy to which performance parameters can be determined and hence verify the "as-built" quality of structures well outweighs (*Figure 2-1*) the common prescriptive approach limitations and assumptions that are: difficult to prove or measure in practice; and do not relate to a service life requirement.

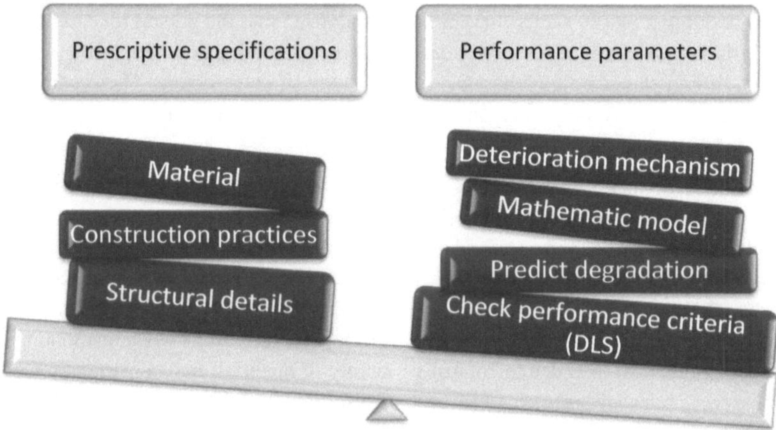

Figure 2-1 Prescriptive specifications vs. Performance parameters (Source: Author)

2.2 Durability of concrete: design codes and standards

Durable concrete depends upon a suitable selection of materials and construction methodologies. However, during construction, the durability of RC structures can be altered in many ways. Some of these are uncontrollable or observational, such as the exposure conditions in the specified environment which is correlated to deterioration rates in degradation models, introducing a certain degree of uncertainty.

Others are defined in more experimental terms which lends itself to testing, such as the durability performance in relation to concrete composition and execution referring to segregation, compaction, curing, bleeding, finishing and micro-cracking. In the first chapter, it was stated that correlation is required between at least two stages indicated *Figure 1-1* as opposed to causation.

Material conditions refer to the mix proportion parameters and quality of individual constituents, whilst manufacturing conditions refer to general production or workmanship during construction. Testing conditions involve sampling, specimen preparation and equipment which are commonly standardised; however, when samples are obtained from the field, the extent of initial curing may differ which inherently introduces additional variability. The field conditioning procedures consider an array of factors that therefore influence test results for the required "performance" concrete.

Whilst numerous studies have embarked on experimental investigations and procedures to assess concrete durability under controlled conditions, the interplay of both experimental and observational characteristics during early-age site conditions has the greatest effect on long-term performance. Field studies that aim to be representative of the entire set of influencing parameters can be closely correlated to as-built performance, and hence their sensitivity to changes during a construction period can provide important details as to identifying potential sources of variability and defining a proper safety margin. It is recognised that the pursuit toward ultimate durability criteria is indeed an iterative one, dependent on optimising and monitoring certain parameters from design stages in the laboratory until the end of construction. These parameters influence concrete durability properties and are well recognised but undoubtedly further refinement is needed regarding their variability.

Kessy (2013) stated the current DI values proposed for evaluating the performance of concrete are based on the short-term monitoring of structures under their actual environmental conditions and can therefore be used in hybrid durability design approaches. DI values must be refined such that they are linked to Service Life Models (SLM's) for performance-based durability specifications. When Performance Based Specifications (PBS) are considered, DI values for CCI and OPI are required to be incorporated in relevant SLM's for estimating DSL. These specifications are intended to control variability of materials and construction methods through measuring relevant properties that account for durability. The design work-flow procedure for both the prescriptive and performance-based durability design approach with corresponding code examples is given in *Table 2-1*.

Table 2-1 Design work-flow procedure for durability design approaches (Source: Author)

	Prescriptive based approach	Performance based approach
Requirements	<ul><li>Intensity of specific Environmental Action is divided into several Qualitative Grades</li><li>The following is then decided upon based on Exposure Environment and Intended Service Life<ul><li>Material Content/Composition</li><li>Construction Practice</li><li>Structure Details</li></ul></li><li>The expected service life is not computed</li></ul>	<ul><li>Select Environmental Action & Deterioration Process</li><li>Quantify Intensity as boundary conditions for model</li><li>Express and Quantify DLS ito Deterioration Effects</li><li>Involved Mechanisms → Mathematical Models</li><li>Numerical Prediction → Evaluate Degradation Extent</li><li>Check Relevant Performance Criteria (Safety Factor or Reliability Index)</li></ul>
Examples	<ul><li>EN (Europe)</li><li>ACI (United States of America)</li><li>CSA (Canada)</li></ul>	<ul><li>DuraCrete (Netherlands)</li><li>*fib* Model Code (Swiss)</li><li>RILEM & BS (United Kingdom)</li></ul>

2.2.1 Environmental exposure classifications

A critical analysis of the prominent features among various environmental exposure classification systems around the world was conducted by Kulkarni (2009). The study emphasised on the limiting values for properties of concrete for various types in specifically prescriptive approaches, as opposed to providing a complete comparison of all durability provisions (performance or model-based approaches). Trends in premature deterioration of RC structures were confirmed resulting in a general use of stringent limiting values for concrete properties. A main finding was that most international standards remained 'prescriptive' in nature, despite an expansion of sub-classes in exposure conditions and alignment with their predicted severity of exposure during service life.

This was done in order to instate a sense of clarity and easily relate typical examples for guidance in design. These sub-classes commonly split the intensity of exposure classifications into several qualitative grades. Taking it a step further would involve defining these classifications into more accurate quantitative grades which is primarily accounted for in Service Life Models (SLMs) in the performance or model-based method. AS 3600 was among one of the first codes to include an extensive definition of coastal, tidal and spray zones. The classification involved defining the coastal zones into three categories (within up to 1 km, beyond 1 - 50 km and beyond 50 km) with corresponding exposure classifications. Tidal or splash and spray zones are defined in terms of 1 m ± highest/lowest astronomical tides and 1 m above wave crest levels, respectively.

A striking difference in ACI 318, is the exclusion of a separate exposure classification for carbonation. Where corrosion protection of reinforcement is necessary, a C1 classification corresponds to carbonation-induced corrosion (no external sources of chlorides), whereas a C2 corresponds to chloride-induced corrosion (external sources from de-icing chemicals, salt, brackish water, seawater or spray from these sources). Prioritising structural elements in marine environments might be problematic according ACI 318, since there is no definition of coastal, tidal or spray zones. In addition, the different external sources can vary quite significantly in magnitude of chloride ions, yet they are all defined under the same severity.

2.2.2 Prescriptive-based specifications

Many of the early signs of concrete deterioration resulting in premature failure and the need for costly repair and rehabilitation of RC structures are the consequence of outdated deemed-to-satisfy rules in prescriptive-based design approaches. Durability is defined as a measure of concrete performance in service, and the ability of concrete to withstand attack by aggressive actions. Hence, limiting values for cement content and water: binder ratio that primarily relate to the compressive strength of concrete have been rendered insufficient in designing for concrete durability, and it is now evident that these restrictions in mix designs are fast approaching their limits of applicability. The backbone of the performance-based methodology does not rely on a single characteristic parameter such as compressive strength, which is far from perfect for predicting concrete durability. The compressive strength of core specimens depends on the slenderness of the specimen, capping material, rate of loading and moisture content which are aspects that are either different, absent or much higher than in real life which do not associate to the transport mechanisms affecting concrete durability, whereas the Durability Index (DI) values do.

These rather traditional approaches consist of AS 3600, BS 8500, ACI 318, CSA A23.1/A23.2 and SANS 10160 which all specify design aids and limiting values for mix design purposes such as maximum w/b ratio, minimum concrete grade and cover depth. EN206-1 is also similar in nature, however, there is an additional provision for minimum cement content conforming to EN 197-1.

A special durability provision exists in AS 3600 that specifies an environmental classification 'U' that refers to an undefined condition in which the degree of severity is unknown. This code, however does not give any guidance on limiting values for concrete composition/proportion, but rather reference should be made to AS 1379 which divides concrete into 'Normal Class' and 'Special Class'. In specifying the required cover based on the characteristic strength of concrete, preference is given to rigid formwork and intense compaction over standard formwork and compaction, in which the later results in much stricter values to be adopted. Characteristic strengths are predefined for different exposure classifications which also greatly depends on the initial continuous curing duration providing an additional requirement for strength upon completion of curing. Prescriptive requirements and limitations for CSA A23.1/A23.2 are very similar to that of AS 3600, however, they are slightly more specific in terms of cover depth including an allowance in design for deviation and increased cover (75 mm) for members cast against and permanently exposed to earth (same as for ACI 318). However, in reality, these limiting values provide little or no indication as to the quality of concrete in relation to its transportation mechanisms.

2.2.3 Performance-based specifications

Concrete durability has proven to be best defined, modelled and tested through the implementation of performance-based specifications, which in general, have exposed limitations in the general rules governing prescriptive-based specifications which depend on the historical performance of the construction industry. Prescriptive-based specifications fail to consider the use of alternative binder systems and ignore the achieved quality of the cover concrete that is responsible for mitigating the ingress and deleterious effects of harmful substances. The framework offered by performance-based specifications allows for the prediction of durability in RC structures based upon the assessment of DI values, monitoring parameters and the use of mathematical models. The success of the DI approach can be accounted to a system of classes linked to various exposure environments in which the compliance of a specific concrete composition is verified through performance tests.

The increased advantages of performance-based specifications are the test methods that allow for the "as-built" quality of the actual structure to be assessed and hence appropriate action can be taken in the case of deviations or non-conformities. In this method, one needs to quantify the environmental action intensity as boundary conditions for the model. Using numerical predictions, the degradation extent is predicted over a given period. The use of tests in performance or model-based methods allow one to provide input parameters such as DI values into Service Life Models (SLM's) which simulate the environmental action according to the defined exposure class and hence relate the degradation for a specific material in relation to the structure itself.

The suite of test methods used in these approaches are central to the transport mechanisms in concrete such as diffusion, migration, permeation, sorption, convection and wick action. Tests are also specifically oriented at the quantity and quality of the concrete cover layer which provides relevance to the actual deterioration mechanisms. CSA A23.1/A23.2 provides performance criteria only in the severe areas for chloride exposure (C-XL or C-1) and chemical attack (A-1) in the form of a maximum imposed limit on chloride ion penetrability measured in coulombs at 56 days' age, similarly to that required in ACI 318.

Both codes also permit a sense of flexibility, such that a variety of cementitious materials may be used to provide concrete of low permeability, specifically in designing for sulphate attack. However, both codes do lack in the sense of providing as such refined limits for acceptance criteria relating to chloride ion penetration (ASTM C1202). The corresponding test for sulphate exposure, ASTM C1012, does however provide strict performance criteria with respect to a maximum expansion at specific time intervals for each relevant exposure class. In terms of freeze-thaw attack, two other performance-based tests can be used, namely, ASTM C666 (rapid repeated freeze-thaw cycles) and ASTM C457 (air-void system determination), which also provides strict performance criteria required for conformity.

2.2.3.1 Example of classification of environmental action type and intensity

A relatively new performance-based approach is given below with reference to the Chinese Model Code. CCES01-2004 defines durability as the ability of a concrete structure to maintain its service performance under environmental actions during its expected service life. The environmental action describes the external solicitation; the expected service life defines the valid duration; and the Service Performance Level is the reference limit state. These aspects are summarised in *Table 2-2*. In design the selection of an appropriate Environmental Class will lead to the classification of an environmental action type and intensity. The steel corrosion process can be accurately understood with identified mechanisms, proposed mathematical models and established monitoring and prevention techniques whereas other processes are not yet at this stage (Li et al., 2008). Failure mechanisms such as the transport of chloride by diffusion in concrete structures that initiate corrosion are not as straight-forward or sequential, nor easy to define and quantify.

Table 2-2 Classification of environmental action type and intensity (Li et al., 2008)

Class	Environment	Intensity	Deterioration process
I	Atmospheric	A,B,C	Carbonation-induced corrosion
II	Freeze-thaw	C,D,E	Internal pore water freezing due to frost
III	Marine	C,D,E,F	Chloride-induced corrosion
IV	De-icing and other salts	C,D,E	Chloride-induced corrosion
V	Chemicals	C,D,E	Industrial polluted air, salt crystallisation or aggressive agents in soil and ground water

CCES01-2004 defines three Durability Limit States (DLS): initiation of the electrochemical process of steel corrosion by a carbonation front transgressing concrete cover or chloride accumulation reaching critical concentration at steel surface; corrosion to an acceptable extent; and concrete damage to an acceptable extent. DLS can be defined at either Serviceability Limti State (SLS) or Ultimate Limit State (ULS). The DLS should be defined in terms of the deterioration process and acceptable extent of deterioration. CCES01-2004 recommends a reliability index of 1.5 for DLS with a failure probability of 6 %, whereas the *fib* Model Code prescribes a reliability index in the same order of 1.8 with a failure probability of 4 %. As you increase the reliability index, a decrease in failure probability is observed, as expected.

DLS defines an acceptable level of deterioration of structural concrete subject to environmental actions. According to CCES01-2004, this limit state belongs to the SLS in conjunction with deformation, crack and fatigue control. The partial safety factors or reliability index which is established at SLS level should be applicable to durability design.

Deterioration processes such as alkali-aggregate reaction, sulphate reaction and concrete surface wearing are considered as special cases and not dealt with in code's environmental action type and intensity classification. CCES01-2004 grades intensity of all environmental action from A to F with increasing severity. One should be cautious when dealing with the "C" intensity for different environment types. Even though the structure has deteriorated to similar conditions, the specific requirements will be based on the environmental action type.

2.2.3.2 Example of Service Life Prediction Model

Schueremans & Gemert (1997) defined a Service Life Prediction Model for RC treated with Water-repellent Compounds based on measured material properties and chloride profiles. Service life prediction can be performed using reliability and stochastic concepts. Reliability analysis is applicable to concrete deterioration associated with steel corrosion initiated by the action of chloride ions. To estimate the service life of a given concrete element, many assumptions must be made that are not valid for concrete. However, to model the chloride transport process in a concrete porous material due to diffusion, it is assumed that Fick's second law applies. Therefore, it is assumed concrete is a homogenous an isotropic material and the medium is non-reactive and non-absorptive. When pores are empty, capillary forces transport the outside solution with chlorides into the concrete. Note the diffusion process is only valid in saturated conditions. Once chlorides reach the reinforcing steel, corrosion begins and delamination/spalling result over a period of time. The following presents a summary of the work done in terms of the diffusion law by Schueremans & Gemert (1997).

- If assumed that no reaction occurs between concrete and free chlorides, an explicit solution of this differential equation can be obtained using the following boundary conditions:

- C = f(x, t=0) = C0 ; 0 < x < ∞ (the initial chloride concentration in the concrete mix)

- C = f(x=0, t) = CS ; 0 < t < ∞ (the chloride concentration loading from the marine environment)

$$C_i(x,t) = C_0 + (C_S - C_0) . erfc\left(D\frac{x}{2\sqrt{Dt}}\right)$$

$$C_i(x,t) = Amount\ of\ chlorides\ at\ (x,t)$$

- A reliability analysis is used to evaluate the probability of failure of the structure or element with a single continuous limit state function g(D). Since only the diffusion coefficient D is random therefore:

   - $g(D) = C_T - C(D)$
   - $Where\ C_T = Threshold\ chloride\ concentration$
   - $And\ C(D) = Chloride\ concentration\ at\ (x,t)$
   - $If\ C(D) < C_T\ then\ g(D) => 0\ ("Safe"\ state)$
   - $If\ C(D) > C_T\ then\ g(D) = < 0\ ("Unsafe"\ state)$
   - $Therefore\ probability\ that\ C_T\ is\ exceeded\ can\ be\ expressed\ as:$
   - $P_f = P(C > C_T) = 1 - F_C(C_T)$
   - $Where\ F_C(C_T) = Cumulative\ Distribution\ of\ C$

      - $Note\ one - by -$
        $one\ relationship\ between\ C\ \&\ D\ so\ exceedance\ probability\ P_f\ can\ be\ rewritten\ as:$
      - $P_f = P(C > C_T) = P(D > D_T) = 1 - F_D(D_T)$

- *Where $F_D(D_T)$ = Cumulative Distribution of D*
- *D_T = Threshold diffusion coefficient obtained from inversion of solution of Fick's law*
- *Where $D_T = f^{-1}(C_T)$*
- *Assuming the diffusion coefficient D has a lognormal distribution,*

  the exceedance probability:

  - $P_f = P(C > C_T) = P(D > D_T) = 1 - \Phi\left(\frac{\ln[D_T] - \lambda_D}{\xi_D}\right) = \Phi\left(-\frac{\ln[D_T] - \lambda_D}{\xi_D}\right)$

- *Where λ_D and ξ_D = Lognormal distribution paramters*
- *And $\Phi(D)$ = Standard normal cumulative distribution function*
- *At increasing depth x (mm) various % Cl, $\frac{Cl}{cem}$ or $\frac{Cl}{H2O}$ = Water soluble chloride content*

- In order to obtain the diffusion coefficient, the following aspects should be considered:

  - C_0 = initial chloride concentration at erection time

  - C_S = chloride concentration loading from the marine environment

  - C_S taken as higher than the concentration of the salt sea-water of 3.5 %

  - Inversion of Fick's second law only soluble when $C_0 < C_i(x,t) < C_S$

  - Least square optimisation proved C_S = 7 % by weight of water (Cl profiles)

  - Note in the tidal zone due to salt crystallisation and the presence of alga at the concrete surface C_S = 9.64 %

- The chloride ingress process can be described in two steps:

- From water in fresh mix:

  - 5 % insoluble salts or locked in a pore of the silicates that are insoluble in water

  - 85 to 90 % soluble salts (Salt of Friedel: C3A.CaCl.10H2O)

  - 5 % free chlorides in solution or easily soluble by adding water

- From marine environment or de-icing salts:

  - Chlorides react very little with the solid phase of concrete and are found as free chlorides whereas soluble salts (Salt of Friedel) acts as a stock of free chlorides to the water in pores

  - Pore water becomes enriched until a final concentration equals product of solubility and the solubility product constant, K_{sp}, is the equilibrium constant for a solid substance dissolving in an aqueous solution. It represents the level at which a solute dissolve in solution. The more soluble a substance is, the higher the K_{sp} value it has

  - Corrosion risk can be attributed to both the chlorides in the pore water (free chlorides) and a part of the soluble chlorides

  - Note for a given amount of chlorides in the pore water, the corrosion risk is higher for a carbonated concrete structure

2.3 Durability of concrete: construction (the reality)

2.3.1 Critique of Durability Provisions

Evidently, various model-based design approach exists successfully worldwide. The validity of the current approach largely depends on the accuracy of the locally used SLM's. Studies have shown that calibration of at least 10 years or more of field data is significant to minimise variability based on predicted and actual or in-situ results (Foster, Stewart, Loo, Ahammed, & Sirivivatnon, 2016). The assessment of the reliability of these design models are of utmost importance. In all cases, the use of poorly calibrated models can either result in over-designed and uneconomic structures or structures that are prone to early failure (either SLS or ULS) that result in catastrophe.

Foster, Stewart, Loo, Ahammed, & Sirivivatnon (2016) conducted the first of a two-part study for the calibration of AS 3600, focussing on the statistical analysis of material properties and model error for the design of beams/slabs in bending and shear columns under combined bending/axial loading. The identification of improvements in concrete and steel reinforcement production had notably reduced variability in material properties. The result was potential to increase code strength reduction factors and eliminate unnecessary conservatism in design. An important point to note in this study was the sample size used that was required in order to eliminate such variability - in total, over 20000 concrete cylinders were statistically analysed from around the country in terms of their strength and variability under standard curing conditions.

A durable concrete structure must start with a durable concrete mix composition and constituents that can withstand the multitude of distress mechanisms that severely affect its service life. This is a task much easier said than done since the primary attributes of concrete degradation include the presence of a gaseous substance (oxygen or carbon dioxide) and water. To eliminate the exposure of cover concrete and the steel reinforcement from such natural sources is impossible, however performance-based specifications allow for the required concrete cover to be designed dependent on the concrete material performance and expected environmental load during service life.

This specific material performance however needs to be achieved in construction regimes and verified through quality control mechanisms for structures to reach their intended target service life. Durability specifications in South Africa for bridges on national roads involve the casting of concrete panels and monitoring of Durability Index (DI) parameters at various stages during construction to establish a correlation to actual in-situ performance. Durability performance is typically evaluated at two stages, these being laboratory and field conditions, which relate to the potential and as-built quality of the structure. Specimens obtained from the field are tested for acceptance purposes since the test results provide a correlation to actual in-situ performance.

From batching of the concrete on site, many aspects can potentially alter the durability properties of the concrete found within the as-built structure. Ensuring formwork is adequately in place, reinforcement steel corresponds with bending schedules and the minimum cover is achieved are all pre-inspection checks, which even if met do not guarantee concrete durability. Some aspects are controllable and can be prevented, however due to the relatively longer time frame required for these defects to become visibly identifiable (months or years into the service life of the structure), there is a high probability of such aspects going unaccounted for. Delays in concrete arrival in hot weather conditions, the further addition of water to concrete to extend its workable life, inadequate compaction in areas of highly congested reinforcement and insufficient curing of exposed surfaces or other faces (upon removal of formwork) are aspects that all compromise concrete durability performance.

Therefore, robust quality control tests are needed to identify instances of such material variability and poor construction practices. To account for material variability and construction error, implementations of "deemed to satisfy" and rigorous approaches often consist of a 'trade-off' between material quality and cover which is currently implemented in the current system of Concrete Durability Target Specifications (Alexander, Ballim, & Kiliswa, 2013).

The implementation of performance-based specifications for concrete durability in South Africa have been on the rise in the contract specifications for diverse and large-scale infrastructure projects in South Africa. These include multi-level interchanges (Umgeni and Mt Edgecombe), bridge widenings (Umdloti River and Tongaat) and the upgrading of national routes (Gauteng Freeway Improvement Programme). The inclusion of these specifications has already offered substantial advantages in both sustainable development and durability of construction. In addition, performance-based specifications can eradicate ineffective quality assurance procedures which are the resultant of most common prescriptive-based specifications and hence decrease risk borne by clients. Over the past decade and since the introduction of performance-based specifications, a much more substantial onus rests on design engineers, concrete producers and contractors in order to promote and encourage innovation in RC structures.

The key elements to be considered in drafting specifications for concrete durability are structural safety, cost, constructability, availability of local materials and laboratories in order to carry out DI tests to the required precision (Kessy et al., 2015). A clear majority of durability specifications from standards in an international context reveal most codes are still prescriptive, with the exception of a few having some performance requirements. One of the broader perspective durability aspects outlined by Kessy, Alexander, & Beushausen (2015) is a client service manual that provides all the necessary information pertaining to the material, manufacturing and testing conditions which can be consulted upon during future maintenance strategies (repair and rehabilitation) and is a prime advantage from a client's perspective.

2.3.2 Quantification of Concrete Variability

Concrete variability is measured by the standard deviation or coefficient of variation (CoV) for compressive strength as well as for durability parameters. The sources of variation attributed to strength hold the same for durability parameters which, broadly stated, arises due to the material, manufacturing and testing conditions (Obla, 2014). These sources ultimately determine the achieved as-built quality and in-situ performance of RC structures which will be expanded upon in Section 2.3.3 Quantification of Concrete Quality. Material variations refer to standards maintained by the concrete producer such as variations in cement, supplementary cementitious materials (SCMs) or additions as denoted in EN206, admixtures and aggregates (both fine and coarse), whilst manufacturing procedures are twofold requiring the responsibility of both the concrete producer and contractor. This refers to variations in the concrete mixture due to proportioning, mixing, transporting and temperature which in turn affect the slump, workability and air content. On the other hand, testing conditions involve sampling, specimen preparation and equipment which are commonly standardised. However, since specimens are obtained from the field for durability parameters, additional variability is encountered through the "extent" of initial curing. During the air-drying process, specimens are exposed to the effects of wind and varying temperature or relative humidity which results in additional variability.

Alexander, Bentur, & Mindess (2017) identified trends that related the variability of concrete quality according to *Figure 2-2* and *Figure 2-3*. This theory is in line with the framework developed by Alexander, Ballim, & Stanish (2008) in order to characterise the durability performance of RC structures by measuring suitable quality parameters representative of the cover layer of laboratory and in-situ concrete. This framework defines the dual aspects of material potential and construction quality, which are important points relating to defining a proper safety margin during construction.

By assuming the same averages for both material potential and "as-built" quality, it is evident that greater variability exists for in-situ values. A means to account for such variability consists of using the test coefficient of variations (COV's). Potential characteristic values obtained under laboratory conditions can hence be 'offset' in order to determine the achievable in-situ or 'as-built' values. These conceptual relationships between material potential and as-built test distributions for a typical DI test with higher values representing better quality were used to create a conceptual framework for the database structure which can be found in Appendix A. The sensitivity of the DI tests is a primary advantage over other prescriptive requirements that do not take into account material factors and construction effects.

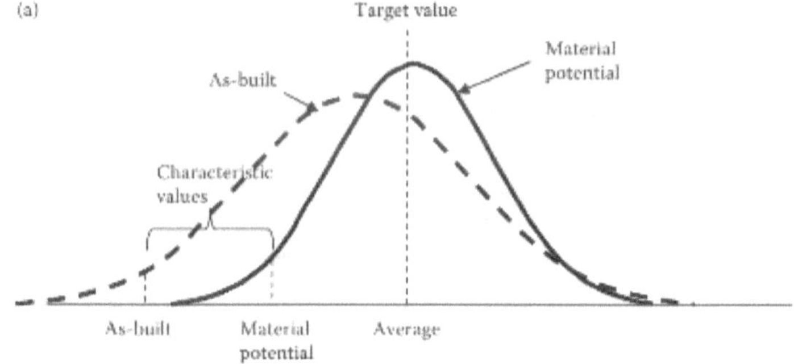

Figure 2-2 Lower average as-built values (Alexander et al., 2017)

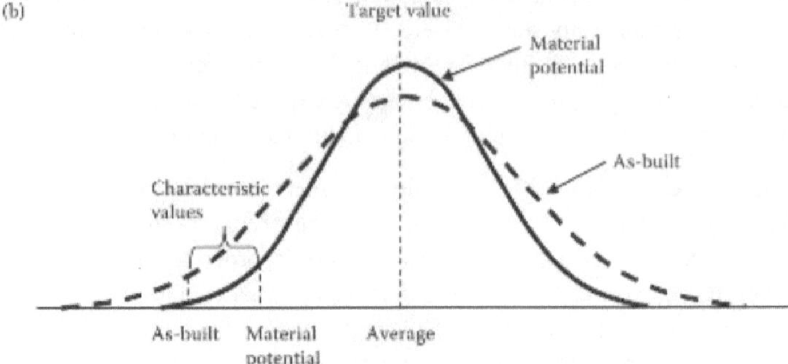

Figure 2-3 Equal average values for material potential and as-built quality
(Alexander, Bentur, & Mindess, 2017)

The output is one of the crucial points of the database and will change depending on what questions we ask the data. Quality control (Scheme 1) in COTO Series 8000 sets out criteria for tests and measurements to control the relevant properties of the "workmanship" and "materials supplied" according to two statistical judgement plans. Both concrete cover and concrete durability are categorised according to Judgement Plan A in which the variability of the test results is not computed, and individual test results are assessed in line with the specified requirements. On the other hand, Judgement plan B is used for in situ densities, strengths of concrete and "certain" other properties. Here the variability of the values of tests is calculated and applied to determine acceptance limits for sample means. Variability in concrete durability properties is encountered that cannot be easily predicted nor quantified.

2.3.3 Quantification of Concrete Quality

The practicality of Durability Index (DI) performance-based specifications to control concrete cover quality was investigated, whereby some of the chief aspects considered involved measuring the extent and magnitude of variability of the test results both within and between projects (Nganga, 2011). The general applicability of the system on construction sites was also investigated i.e. through core extraction from test panels and the use of local laboratories to execute the test methods.

These within and between CoV studied by Nganga (2011) give an indication of the repeatability and reproducibility of the test results for which predetermined levels of precision are defined as in (Stanish, Alexander, & Ballim, 2006). In the earlier study, an inter-laboratory test scheme was conducted to confidently measure the repeatability (single operator CoV) and reproducibility (between laboratory CoV) of the DI tests. These measures provide important information in order to specify limiting test values to obtain the required performance. A similar phenomenon can be drawn to compressive strength, in which both target and characteristic values are specified in order to account for variability.

For Oxygen Permeability Index (OPI), it was found that the CoV's were approximately the same within laboratory and between laboratories, therefore the variation cannot be accounted to the test methods but rather to the inherent material variability. The OPI test is sensitive to the compaction degree and it is expected that this will vary more than the proportions within a particular batch. For Water Sorptivity Index (WSI), literature suggests the results are insensitive to variations in strength and composition, which is consistent with this study. WSI is more significantly affected by early- age ($\leq$ 7 days) curing conditions, with these variations diminishing greatly following longer water curing periods.

For Chloride Conductivity Index (CCI), a large number of the results had to be eliminated due to high variability. This was mainly attributed to improper sealing and incomplete saturation of the specimens that resulted in high and low values, respectively. The former resulted due to equipment alterations in the size of the core barrel, a difference of precisely 2 mm. Considering the tight tolerance on the test rig, equivalent to this marginal difference, laboratories did experience problems when samples were either smaller or bigger than the core barrel. A subsequent recommendation from this study was that the CCI test apparatus be redesigned in order to increase the tolerances.

The extent and magnitude of variability outlined by Nganga (2011) means that despite resultant average DI values passing the specification limit in some projects, alarming amounts of defectives are still present. In this regard, high proportions of defectives has to be accounted for in specifications. The contract specification from the client need to define the desired level of performance, specify the frequency of testing, set out the limits for acceptability and define conformity rules linking to action (acceptance or contractual penalties / remedial action). Therefore, the roles and responsibilities from project identification to site handover are in need of a change of mindset for all the stakeholders involved in ensuring concrete durability (Kessy, Alexander, & Beushausen, 2015).

Before construction and testing, the contractor needs to ensure that the pre-qualification tests conducted by the concrete producer can be verified i.e. the fresh concrete can be transported from the discharge point and maintains the desired quality in its hardened state after accounting for construction practices and field variability factors (Kessy, Alexander, & Beushausen, 2015). During construction and testing, the clients' representative should be able to verify that the durability requirements contained in the specifications have or will be satisfied during the contact. This is ultimately where the conformity rules can be consulted upon and compliance can be measured.

As suggested by Alexander, Ballim, & Stanish (2008), a 1:10 chance should be adopted at this stage for the DI tests. This is indicative of a 90 % confidence level and corresponds to an approximate safety margin of 0.3 (log scale) below for OPI and 0.2 mS/cm above for CCI and 1.0 mm/hr$^{0.5}$ above for WSI. Compressive strength is an ultimate limit state (ULS) criterion and hence the characteristic value is set where 5 % (1:20) of the total area under the curve falls. In other words, the characteristic strength is defined as the strength of concrete below which not more than 5 % of the results are expected to fail. On the other hand, durability is defined as a serviceability limit state (SLS) criterion and hence the 95 % confidence level is too strict for application and should be lowered as indicated. By analysing data from a multitude of different construction projects representing a variety of material, manufacturing and testing conditions, these proposed safety margins which are split according to acceptance and rejection limits can be evaluated against project specific DI test results to quantify concrete variability with regard to durability performance on construction sites.

According to current trends of OPI results compared by Nganga, Alexander, & Beushausen (2017) from two different periods. The first being at the introduction of the performance-based approach on a full-scale level (2009 – 2010) and the second concerned with the increased implementation and more current available data (2011 – 2015). Increasing trends in the variability of test results over the years provide a major concern. The latter period displayed on average lower OPI results, a higher Coefficient of Variance (CoV) and an increased number of defective units below the threshold OPI value (9.40) which are all indications of poor, variable or ineffective manufactured quality (compaction & mix design) and curing.

Semi-invasive testing has allowed for the development of performance-based durability design specifications in South Africa which consist of the coring of trial panels that are cured on site as a mechanism to ensure quality control for durability concrete. Studies conducted by Ronny (2011) have in addition corroborated the durability results from these trial panels with that measured from in-situ cores. Despite providing slightly superior results as indicated in the below figures, at present, this is the most feasible means to replicate the material and manufacturing conditions and assess results in a desirable and non-destructive manner.

Ronny (2011) tested the following hypothesis: Coring of trial panels and/or test cubes cured on site will replicate results from cores drilled from the structure and therefore can be used to replicate durability. It was found that the effects of a confined space combined with a controlled curing environment were more pronounced on cores extracted from test cubes resulting in much superior values than those for durability panels and the in-situ concrete. It was stated that durability panels sufficiently replicate the durability of the in-situ concrete due to the common exposure environment, curing, placing and compaction methods, however, the DI results from different mix designs and projects reveal that trial panels contained superior results than in-situ cores on 4 out of 5 occasions for both OPI and WSI.

In *Figure 2-4* and *Figure 2-5*, the general trend observed in DI values is as expected, with the in-situ concrete displaying the lowest (for OPI) and highest (for WSI) values for either parameter. The trend occurs for both mix designs, except on one occasion for trial panels cast in the field containing superior results than cubes cast in the laboratory which although is controversial, is repeated in the OPI and WSI results. In this sense, the higher OPI combined with lower WSI associate well and increase the reliability of the results even with the lower correlation coefficients in *Figure 2-6* and *Figure 2-7*. However, it is known for the Black Mfolozi River Bridge contract that the project specifications opted for labour intensive operations for all concrete as opposed to the general plant intensive methods (Ronny, 2011). Furthermore, such occasional reversals have been known to occur as result of good curing and construction practices for example, considerable densification to the surface of well-cured ground slabs (Alexander, Ballim, & Stanish, 2008).

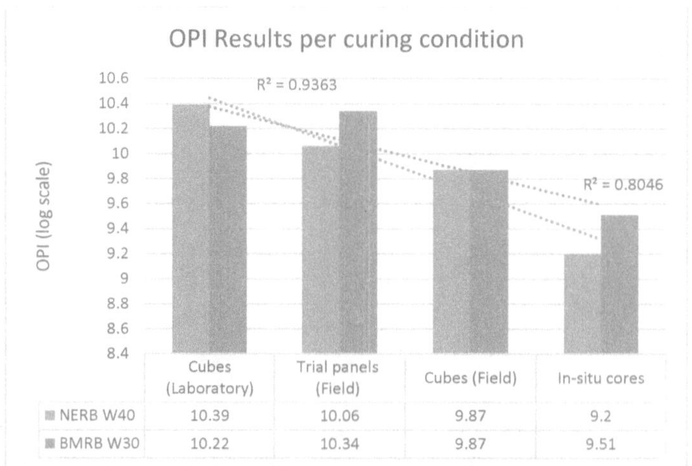

	Cubes (Laboratory)	Trial panels (Field)	Cubes (Field)	In-situ cores
NERB W40	10.39	10.06	9.87	9.2
BMRB W30	10.22	10.34	9.87	9.51

Figure 2-4 OPI Results for New England & Black Mfolozi Road Bridges (Ronny, 2011)

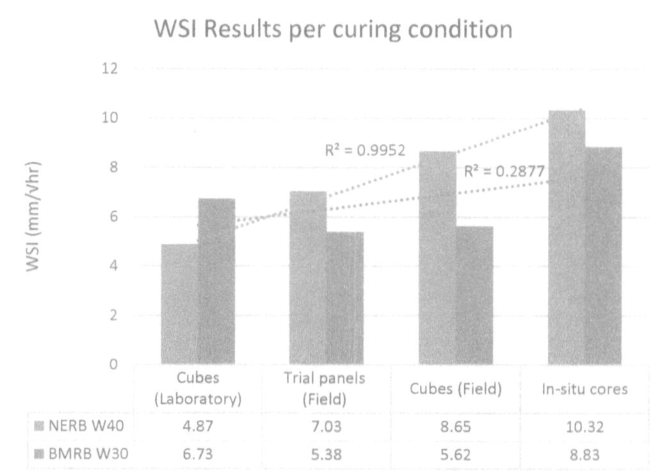

	Cubes (Laboratory)	Trial panels (Field)	Cubes (Field)	In-situ cores
NERB W40	4.87	7.03	8.65	10.32
BMRB W30	6.73	5.38	5.62	8.83

Figure 2-5 WSI Results for New England & Black Mfolozi Road Bridges (Ronny, 2011)

It was found that test cubes cast under laboratory conditions are ineffective to predict the durability of the in-situ concrete, however, the DI results from different mix designs and projects reveal cubes cast in the field in wet (submerged) curing conditions also contained superior results than the laboratory conditions on 2 out of 3 conditions for both OPI and WSI. On one occasion, for OPI, the air-cured field specimens contained superior results that the wet-cured field specimens which is although is controversial contains a negative correlation coefficient of 0.7605. However, both these results were lower than the cubes cast under laboratory conditions which indicates that there is a difference between all 3 curing conditions. Nevertheless, the use of test cubes exposed to air and a wet-curing environment can be used on construction sites to display the extremities of curing conditions (Surana, Pillai, & Santhanam, 2017) which will be discussed further in Section 2.3.5 (Quantification of Curing Effectiveness) for five different parameters.

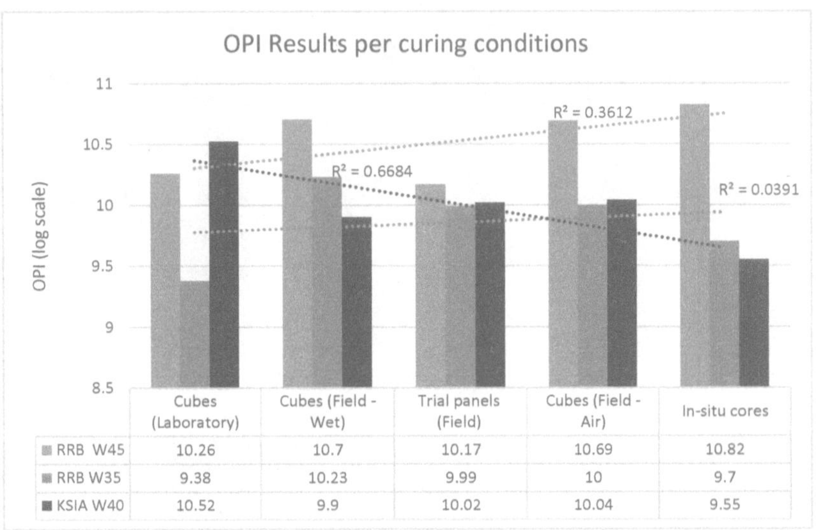

	Cubes (Laboratory)	Cubes (Field - Wet)	Trial panels (Field)	Cubes (Field - Air)	In-situ cores
RRB W45	10.26	10.7	10.17	10.69	10.82
RRB W35	9.38	10.23	9.99	10	9.7
KSIA W40	10.52	9.9	10.02	10.04	9.55

Figure 2-6 OPI Results for Richmond Road & King Shaka International Airport Bridges (Ronny, 2011)

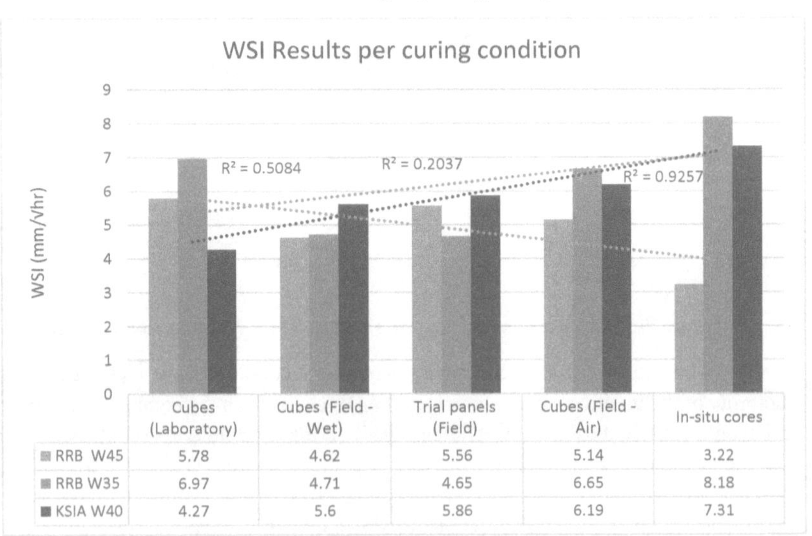

	Cubes (Laboratory)	Cubes (Field - Wet)	Trial panels (Field)	Cubes (Field - Air)	In-situ cores
RRB W45	5.78	4.62	5.56	5.14	3.22
RRB W35	6.97	4.71	4.65	6.65	8.18
KSIA W40	4.27	5.6	5.86	6.19	7.31

Figure 2-7 WSI Results for Richmond Road & King Shaka International Airport Bridges (Ronny, 2011)

Test panels which are rather small in comparison to the actual element being cast, are relatively easier to cure, compact and maintain in a stable environment whilst the concrete hydration reaction is ongoing. However, studies conducted by Gouws, Alexander, & Maritz (2001) support the use of DI values to directly or indirectly control and assess the quality of concrete on site. Research does however indicate that the best way of assessing the performance of placed concrete is by extracting cores from the structure in order to ensure the durability parameter is met. Despite the destructive and often undesirable means, by undertaking this process, the following advantages can be gained from the cores extracted:

- Results from in-situ cores can be checked against laboratory results

- Extent of curing can be determined (if curing parameter is measurable)

- Degree of compaction can be determined and compared

2.3.4 Falsework and formwork

Prolonged and conventional wet curing periods for concrete structures are often impractical due to the constraints faced in construction. Resultantly, concrete protection and curing methods such as the duration of retention for formwork and application of impermeable membranes to prevent water loss have developed, which at large determine the microstructure development characteristics near the cover layer. The global shortage of water and construction related issues that compromise curing performance, which combined with unreliable test methods to evaluate compliance with construction specifications, obstruct the aim of site quality control systems.

A distinction needs to be drawn between the specimen curing conditions and the exposure conditions which can be supplemented with more than one period in the case of field cured specimens. Visser & Han (2003) state that it is sometimes not clear where a curing period will end and where an exposure period will start and for this reason divisions between curing and exposure must be made on the basis of experiments. For curing conditions, a minimum of two curing periods should be the minimum required input for the database i.e. batching (covered) $\leq$ 1 day and submerged (in water) or outdoors (sheltered / unsheltered) $\approx$ 28 days, as and when applicable. For exposure conditions, the main difference to the above is that certain aggressive exposure agents are applicable that can be specified with a corresponding concentration and unit.

Curing conditions can include a continuous 28-day wet curing duration in saturated lime solution or a 7-day period followed by air drying for the remaining 21 days. The latter period is typical of common construction practice assuming that formwork is retained in place for a minimum of 7 days and the materials used comply with thermal insulation and moisture absorption specifications. Another curing condition for specimens can include air drying for the entire duration of 28 days. The distinction to be drawn between the wet curing and air-drying regimes is that these two conditions represent the ultimate extremities of manufactured quality (compaction & mix design) and curing (Surana, Pillai, & Santhanam, 2017).

National road specifications used in South Africa that govern the minimum period in days for the removal of falsework and formwork consist of Table 6206/1 in Committee of Land Transport Officials Standard Specifications (COLTO, 1998). The misleading nature and fact that these specifications only rely on strength, allow shorter periods to be sufficient, if the contractor proves this to the satisfaction of the engineer. However, on construction sites, this is primarily based on the crushing compressive strength ($\geq$ 7 days) of a cube cast using the same mix, which despite being field cured, has no relation to the microstructure development of the cover concrete pore structure. In pre-stressed concrete structures, the problem is further exacerbated, as the superstructure or bridge deck must reach a required minimum compressive strength before stressing can occur, which is advantageous to minimise from a programming perspective. The early completion of pre-stressing works results in early removal of falsework and formwork – a cost, time and access advantage for contractor's dependent on the time period shortened.

It should be noted that the relationship between curing and the physical development of concrete has a strong link to strength, but the greatest variation due to curing will impact on concrete near surface properties and not bulk properties. Tests proven in ACI 308 strongly correlate strength gain for both moist-air ($\geq$ 7 days) and continuously moist (28 day) cured conditions. As result, in special conditions to these standard specifications, it was found necessary to limit the removal of falsework and formwork to a minimum of 7 days, only if retainment of formwork is the only method to cure concrete. However, with the introduction of impermeable membranes, seldom is this the case. Cather (1994) suggested that in order to stress the importance of curing it should be made into a separately billed item in the pricing schedule for the project, which has seen nationwide implementation in most, if not all South Africa's projects.

A literature review by Mekiso (2013) investigated concrete curing and its practice in South Africa which noted the following. In general, curing was not as closely supervised and controlled as compared to batching and mixing operations for concrete. Ensuring that formwork of newly cast concrete structures is in place for $\geq$ 5 days would positively impact on quality in terms of strength and durability. Curing practices which involve the application of water to a certain extent will also sustain hydration and pozzolanic reactions.

Special conditions further stipulate the unprotected concrete can only be left exposed for a maximum of two hours, before impermeable membranes are to be installed as per the manufacturer's instructions. The issue here is that strength gain does not provide the necessary indicator that curing has been undertaken correctly, but DI tests specifically can relate these properties in terms of actual performance from surface properties linked to transport mechanisms occurring at the cover layer. Different types of curing compounds include acrylic resins, wax and resin emulsions in water which are applied to the surface of exposed RC structures upon removal of falsework and formwork. Therefore, the type of element cored (trial or test panel) should relate to the formwork and curing regime experienced by the in-situ structure, should cores not be extracted from it, to correlate to the achieved as-built quality or in-situ durability performance.

In the event of coring the structure, complications arise such as identifying which areas of the structure can be cored without compromising structural integrity, and at the same time, are representative of construction quality. The directions of proposed drilling for cores from vertically and horizontally cast panels must be selected with emphasis on examining certain construction practices (curing, compaction, bleeding, micro-cracking, segregation etc.) for microstructure defects and other phenomenon. Bleeding lenses will form horizontally in *Figure 2-8* and therefore coring at right angles to the casting direction will ensure these construction effects are in the associated test direction in relation to the concrete deterioration transport mechanism. This will ensure the severity of the defect is being measured with such test specimens and the results are as far as possible representative of the achieved as-built quality or in-situ durability performance.

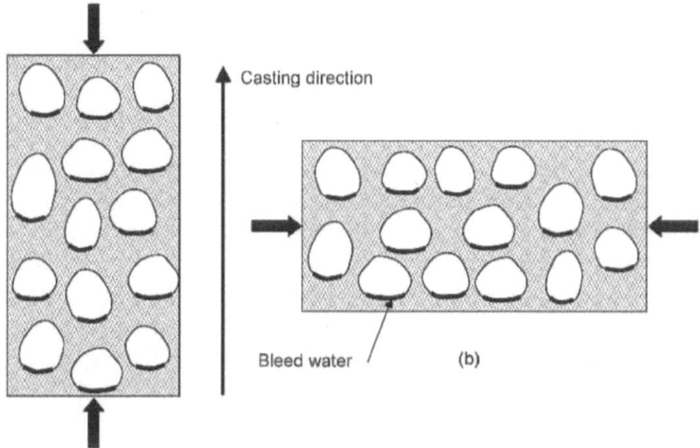

Figure 2-8 Planes of weakness due to bleeding: (a) Axis of specimen vertical and (b) axis of specimen horizontal (Ozyildirim & Carino, 2006)

2.3.5 Quantification of Curing Effectiveness

The effect of curing methods was evaluated on mechanical properties relating to concrete strength (compressive) and durability (water-absorption and chloride permeability) using Portland Cement (PC) and Silica Fume Cement (SFC). The two methods implemented consisted of covering with a wet burlap and/or the application of curing compounds. The latter consisted of coal tar epoxy and conventional water, acrylic and bitumen-based products. The strength and durability results indicated that curing compounds applied without the initial period of wet burlap curing performed similar or better. Significant changes in compressive strength could not be identified according to the selected curing methodology, however this could be done for durability parameters. It was identified that the initial period of wet curing prior to curing compound application increases durability parameters proportionally, with subtle increases noticed from as little as a one-day variation (Ibrahim, Shameem, Al-Mehthel, & Maslehuddin, 2013).

In a further study conducted by Surana, Pillai, & Santhanam (2017), a series of different test methods for durability parameters were evaluated to determine their suitability to characterise and qualify curing compounds for concrete mortar. These methods were compared to the rather conventional approach, which has primarily been to assess curing effectiveness based on field cured compressive strength gain which is the subject of numerous debates. However, a vast number of construction specifications found in Canada, United States of America, Europe and even South Africa still relate curing effectiveness to the field achieved compressive strength gain as previously discussed.

It was proven that at least three of the durability parameter test methods were far more sensitive than compressive strength to detect changes in manufactured quality (compaction & mix design) and curing with OPI subsequently recommended as the most suitable test due to its high sensitivity and general consistency in results for the different curing methods. The other concrete durability parameters investigated involved WSI, non-steady state migration coefficient for chloride penetration (D_{nssm}) and water-penetrable porosity in which the sensitivity of the tests were evaluated in relation to different environments and curing regimes. The study called for further field studies using these suggested methods in order to develop guidelines or performance specifications for the selection of curing compounds (Surana, Pillai, & Santhanam, 2017).

The basis of the study focussed on the absolute extremities of manufactured quality (compaction & mix design) and curing through air and wet curing whereby the performance of curing compounds were assessed in relation to this. Subtle differences were noticed when characterising the durability parameter properties from concrete whereas for compressive strength, the reduced variation and insensitivity to detect changes in manufactured quality (compaction & mix design) and curing inhibits its use as a quantifiable concrete durability parameter. For all five test parameters, the results were expressed in a relative manner with respect to that achieved for 28 days wet curing conditions to facilitate comparison and interpretation. The results obtained from the experimental investigation to determine the suitability of test methods to assess the efficiency of curing compounds is summarised in *Table 2-3*.

Table 2-3 Maximum variability and sensitivity of parameters (Surana, Pillai, & Santhanam, 2017)

Parameter	Mild conditions (25 °C, 65 % RH)	Hot conditions (45 °C, 55 % RH)
Compressive Strength (MPa)	20 %	40 %
Water-penetrable (total) porosity (%)	Insensitive to curing conditions	
Oxygen permeability index (OPI – log scale)	76 %	166 %
Water sorptivity index (WSI – mm/hr$^{0.5}$)	28 %	96 %
Non-steady-state migration coefficient for chloride penetration (D_{nssm})	122 %	158 %

Compressive strength showed the least variations after water-penetrable (total) porosity and hence failed to differentiate between the curing compounds from that of air drying or no curing, however the durability parameters displayed much greater variation and sensitivity to detect changes in curing. Therefore, the subtle differences in trends of durability parameters indicate that curing affects the transport mechanisms in concrete in different ways. Strength and durability properties diminish in the absence of wet curing and an increase in temperature from 25 °C to 45 °C, however curing compounds are designed to resist moisture loss and should improve the performance of concrete properties over air drying consistently. The results for WSI and D_{nssm} displayed lower consistency and increased variation with some curing compounds performing worse than air drying which in turn reduces their reliability as a durability parameter. In addition, water-penetrable porosity was found to be insensitive to curing as the test failed to differentiate between curing compounds, air drying and wet curing.

The importance of the DI parameters over compressive strength is its link to curing through the higher sensitivity and consistency within results for different curing conditions. Existing test methods currently in our specifications (ASTM C156, ASTM C309) for RC structures and concrete pavements have been reported to have large variability, yet still they find inclusion in our performance specifications with reference to the relevant ASTM standards.

The problem with existing methods such as ASTM C156 which is a simple water loss test used for the selection of curing compounds is that it exhibits large variability and has hence been the subject of worldwide critique. ASTM C309 further recommends a limit of 0.55 kg/m^2 for water loss; however, ASTM C156 reports repeatability and reproducibility variability of 0.13 kg/m^2 and 0.30 kg/m^2. This low level of precision results in an undesirable safety margin for the test which undermines its ability to characterise or differentiate between the efficiency of curing compounds. The integrity of the limiting value is also put into question which makes it difficult to decide proper acceptance criteria. DI parameters display greater sensitivity and consistency in results should therefore be used to assess the effectiveness of curing compounds on construction sites for RC structures and concrete pavements. There is a need for these tests to be conducted at relatively early-age ($\pm$ 28 days) and is the case with the current DI tests which further supports its application for quality control mechanisms on construction sites.

To use DI values to assess the effectiveness of curing compounds, adequate limits combined with better control is required for the variability encountered that depends on its source. As mentioned in Section 2.3.2 (Quantification of Concrete Variability), the sources of variation attributed to strength hold the same for durability parameters which, broadly stated, arises due to the material, manufacturing and testing conditions which ultimately determine the achieved as-built quality and in-situ performance of RC structures. In addition to the effects of wind and varying temperature or relative humidity on the variability of DI results obtained from the field, DI parameters assess the actual concrete composition or mix design as opposed to test methods such as ASTM C156 which suggests the use of mortar. Although this facilitates more sensitivity and easier assessment of curing effectiveness, evidently cement additions such as SCMs and coarse aggregates contained in the concrete mix design introduces additional variability which should be accounted for when conducting studies on actual field-cured specimens to develop guidelines for acceptance criteria for the selection of curing compounds.

Coarse aggregates affect the permeability of concrete depending on its type and gradation and a clear majority of structural concrete mix designs contain blended cement which refine the pore structure and increase the resistance to penetration of aggressive agents. However, resultantly the concrete has less capacity to bind CO_2 which is one of the most important factors for carbonation resistance. The transport of CO_2 through a carbonated layer is a secondary deterioration mechanism, hence good curing can partially offset the effects of such lower binding capacity. However, the hydration reactions of blended cements are much slower than those compared to plain Portland cements, thus the beneficial effect of increased resistance can only be achieved if the early-age curing conditions of concrete are adequate.

Even though blended cements have the potential to reach a less permeable state than Portland cements when well cured (under laboratory conditions), the effect of reduced permeability is near diminished in a dry curing environment (under site conditions). Commonly the trend is to specify low w/b concretes since they are less adversely affected by poor curing which is a common occurrence in concrete mix designs, however the practice of good curing should be maintained in the three different and distinct stages in order to correctly interpret the variability. Durability properties should also be used and tested under laboratory conditions to assess the impact of field conditions on the concrete found within the as-built structure. Experimental studies conducted infer that the inner mortar did not show any significant improvement to the outer mortar which is against general expectation for actual structures. This trend is not evident in durability parameters such as water-penetrable (total) porosity and WSI as identified by Surana, Pillai, & Santhanam (2017) due to the high surface-to-volume ratio of a cube that results in high rates of water loss due to the initial porosity of the mortar. The rise of water to the surface would have caused continual loss of water throughout the specimen and hence similarity in results for the inner and outer mortar. It is also well known that concrete cubes, due to relative ease of placing, compaction and curing result in overestimated values to that achieved in actual structures. Despite keeping the concrete mix design and exposure conditions consistent, this superiority is still evident because of such a confined space (the cube) in relation to the structure itself. However, each average DI test result contains up to four individual determinations, therefore, upon further examination of the OPI data in *Figure 2-9*, it is possible to establish the influence of curing with depth even though specimens are obtained from concrete cubes with the following convention in *Figure 2-10*.

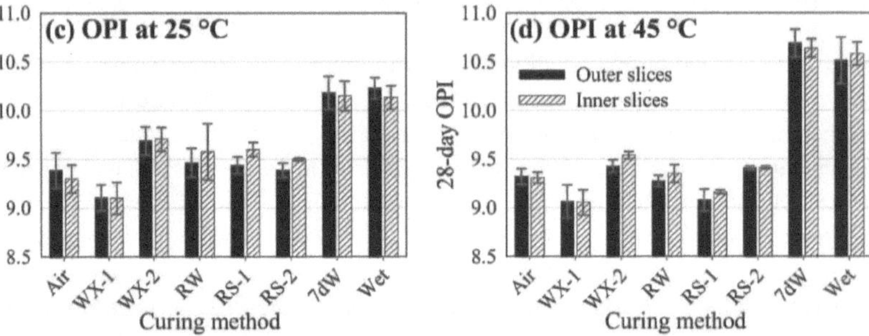

Figure 2-9 Effects of laboratory curing in properties of the near-surface and inner mortar (outer and inner slices) at 25°C and 45°C (Surana, Pillai, & Santhanam, 2017)

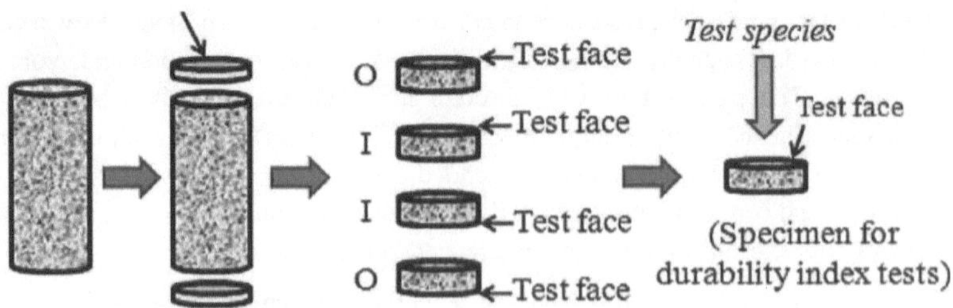

Figure 2-10 Preparation of test specimens for durability tests (Surana, Pillai, & Santhanam, 2017)

At both temperatures, considering air drying and wet curing, all inner slices contain a lower value when compared to the outer slices, indicative of lower quality. However, for curing compounds, such as acrylic resins (RS-1 and RS-2), wax (WX-1 and WX-2) and resin (RW) emulsions in water, the trend reverses, with OPI values for the outer slices either equal to or greater than the inner slices. The effect of curing compounds on the outer surface is clear evidence of its influence on the microstructure in this vicinity of the concrete.

It is also evident that durability and performance of concrete in service is greatly affected by the extent of initial curing conditions encountered in the field. Therefore, it is important to conduct actual field studies to confirm the observed trends. During field conditions, the rate of water loss would be substantially different to what is experienced during the experimental procedure which involved de-moulding of specimens after one day followed by placement into a controlled environmental chamber which result in additional variability. However, field cured specimens ideally replicate the formwork regime of the in-situ structure and are contained in the specified environment under actual exposure conditions, hence are expected to provide an accurate correlation to the achieved as-built quality or in-situ durability performance such as concrete panels of 400 x 600 x 150 mm dimensions as is currently in the South African durability specifications as implemented by SANRAL.

2.3.6 Cover depth

In the Australian Standards (AS 3600 – Section 4), durability provisions were introduced in 1998 which was aligned with the bridge design code (AS 5100.5). This occurred after the publication of numerous reports that indicated increasing signs of distress in older structures were attributed to poor detailing and workmanship or supervision during construction resulting in greater maintenance costs. Stricter tolerances in the construction of RC structures create a smaller margin for error. As result, simple errors arising in either detailing or steel-fixing can easily result in cover deficient zones that are not clearly visible in heavily congested reinforcement areas. Cover deficiencies are probably most influential in reducing the service life of concrete structures as corrosion initiation, which is a function of reinforcement depth, occurs much quicker as result of either chloride diffusion or carbonation-induced corrosion.

Often interpretation from the design engineer is required to identify critical locations in structures that are subjected to the full climatic effects. These locations could consist of the top of abutment/pier faces or the bottom of deck beams/girders situated below bridge expansion joints. Generally, if the face of any structural element with a cover deficiency is exposed to continuous wetting or drying cycles, there is an increased risk of spalling and delamination early during its design service life, dependent on the quality of the cover layer. A proper understanding of the structure's drainage requirements might be necessary in order to understand the various ways that different structural elements are exposed to wetting and drying cycles.

Evidently, the locations in which large areas or numerous bars cause a cover deficiency or irregularity should be prioritised in the Bridge Management System (BMS) and assessed during inspections in order to rate or repair these defects. Merretz (2010) also stressed an equal importance in both specifying the correct concrete cover and the attaining of such cover during construction for structures to remain serviceable and maintenance-free throughout its design life.

Therefore, the recording of the concrete cover for structural elements must form an integral part of the acceptance process for quality control on site which is currently adopted in practice by organisations like SANRAL. Merretz (2010) further identified that in order to successfully achieve capillary discontinuity and limit the ingress of carbon dioxide, oxygen, water and harmful ions over time, cover concrete must also be adequately compacted and cured to ensure it begins its' life as a crack-free medium.

2.3.7 Compaction and curing

Reinforcement inserted into structural elements can consume up to 10 % of the total concrete volume. Furthermore, pre-stressed concrete applications demand high-strength concrete (up to 50 MPa) in order to minimise cross-sectional area and maximise strength, with a common misconception that durability will be accounted for. Over the past decade, it has been proven that no correlation exists between concrete durability and strength, as confirmed by a recent study conducted by Nganga (2011). Specifying a high cement content ($\geq 400 - 450$ kg/m^3) does not guarantee a concrete of improved DI values, if other material factors and construction practices are ignored. The impact of inadequate curing alone can have a distinct effect on durability gain, particularly in the cover region of structures. Inadequate compaction until the faces of formwork in heavily congested reinforcement regions further aggravate this effect which in turn results in a cover region with a permanently defective microstructure (*Figure 2-11*). Subsequent effects are either form of induced corrosion, well before structures reach their design life, especially in severe environments.

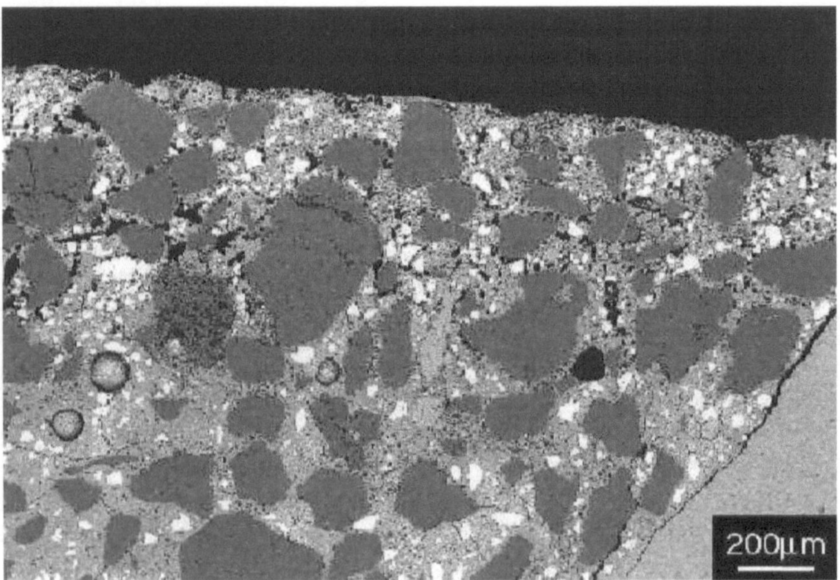

Figure 2-11 SEM Shot of Bridge Deck Core (Poorly cured concrete)
(WHD MIcroanalysis Consultants Ltd, 2005)

2.4 Quality Control Scheme for Concrete Durability

A particular level of an engineering parameter must be defined as an acceptable quality in quality assurance specifications such as COTO (2018b). The goal of any sampling or judgement plan is to distinguish good lots from bad lots where observations may be attribute or variable. The kind of data to be analysed will determine the applicability of either an attribute sampling plan or variable sampling plan. Since DI results are measured on a numerical scale, the variable sampling plan is appropriate, however this contrasts with the judgement plan proposed by COTO (2018b). COTO (2018b) defines two judgement plans for the assessment of test results. Judgement Plan B is used to assess parameters such as the relative compaction and compressive strength which computes the variability of the values from tests which is applied to determine acceptance and rejection limits. Judgement Plan A is used for judging measurement of concrete cover and concrete durability, where it is stated that the number of test results do not allow the use of normal statistical methods. In further accordance with this plan, the compliance of the individual results only with the specified requirements is determined and the variability of test results is not computed.

This statement can be conflicting, since if there is great deviation between results within a project to the specified requirements, the variability is key to interpreting and analysing the data accurately. The variability and acceptance limits of the DI values are best illustrated using probability distributions that describe the data. Through using this distribution in order to analyse results, inferences can also be made based on probability statements for evaluating compliance with the specification according to various limits (acceptance or rejection). It should be noted that COTO (2018b) does not allow for such probability statements to determine compliance.

If the mean of the result is the only specified method of assessment and the variability of test results is not computed, the DI data becomes more attribute than variable. Variable data contains more information than attribute data since it allows an assessment of how poor or good the data is rather than simply assessing whether the lot is defective or not and is therefore directly related to the information from different projects (laboratories and contractors), materials (concrete composition), execution (linked to curing) and environment (linked to exposure).

In applying sampling by variables, an acceptable lot quality can be defined with respect to an upper or lower specification limit. With this boundary condition, the acceptable quality level can also be defined as a maximum allowable fraction of defectives. The boundary condition also referred to as the specification limit (Ls) is the limit value of the property of any product, outside which not more than a certain specified percentage, phi (Ø) of the population of values representing an acceptable product property can lie (COTO, 2018b). The specification limit may be a single lower limit Ls (OPI) or single upper limit L's (WSI or CCI) also referred to as nominal DI values (COTO, 2018a). However, the specification limit is not linked to a maximum specified percentage, phi (Ø) which is further discussed in Section 2.4.5 (Justifying a Maximum Variability or Percentage Defectives).

The DI parameters are assessed in terms of a sample mean ($\bar{X}n$) which is the arithmetic mean of a set of 4 test determinations that constitute the sample. In order to compute the sample mean or sample standard deviation, a minimum number of DI tests are required which are usually sufficient to conduct an outlier test and remove one outlier. The sample standard deviation (S) is the difference between values of an individual sample and the sample mean divided by the sample size. The Coefficient of Variation (CoV) expressed as a percentage is equivalent to the sample standard deviation (S) divided by the sample mean ($\bar{X}n$). COTO (2018b) requires a minimum of four DI test results (16 disc specimens) representing four average results of four specimens each but the number may vary according to the size of the pour or number of concreting days which impact on the amount of trial or test panels cast.

For performance-based specifications to be successful in monitoring concrete durability targets, correlations need to be established between the testing at two or more stages. A vast majority of the Durability Index (DI) results are from test panels which inadvertently bring additional sources of variability, hence without the laboratory cured trial panels or cores extracted from the actual structure, the curing effectiveness and extent of curing in the actual structure cannot be determined. Therefore, the additional two stages which include laboratory and field (in-situ) results can be used to identify unacceptable margins in the test results in order to advise on actual performance on construction sites and future long-term monitoring. The testing at the various stages during construction in order to identify occurrences of inadequate durability (as reflected in the relevant durability index values) is pivotal in order to correctly analyse the data available.

2.4.1 COTO Concrete Durability Specification

The durability specification used in the construction of national road infrastructure is deterministic in nature as opposed to probabilistic or stochastic. A deterministic model is essentially a formula whereby if the starting conditions do not vary; the result can be fairly predicted or assumed. However, if there is deviation in the starting conditions i.e. the DI results in comparison to the specification, then these values must be rechecked to verify the initial design assumptions and ideally quantify the loss of serviceability using SLMs. Stochastic models incorporate one or more probabilistic elements into the model and as such the final output consists of statements based on confidence intervals.

Stochastic models are primarily used to accurately portray the likelihood of an event or series of events occurring. Uses of stochastic models involve risk management and mitigation whereby risk models are expressed as the product of the probability of an event and the cost of the event. Deterministic models are easier to analyse, whilst stochastic models tend to be more realistic, especially in the case of small samples. The criteria to be used in the adjudication of concrete quality has been expanded in COTO (2018b) into five sections, namely laboratory, full acceptance, conditional acceptance, remedial acceptance and rejection, for each of the DI tests indicated in *Table 2-4*. For laboratory, target values must be achieved in wet curing environments. As previously stated in Chapter 1, by also assessing values exposed to air (dry environment), the extremities of construction quality can be gauged, which can advise on the safety margins to adopt during a construction period for durability parameters.

Therefore, this method can be used to characterise and qualify curing compounds n curing specifications. In COTO (2018a), the curing period can be prescribed by a minimum strength or curing efficiency linked to WSI which should be stated on the drawings. Despite the controversial use of WSI as opposed to OPI, in the case of the latter, further there is no guidance is given on how to specify appropriate limits for the parameter. It should be noted that the former minimum strength provisions arise from those discussed in Section 2.3.4 (Falsework and Formwork).

Table 2-4 Criteria to be used in the adjudication of concrete quality (Table A20.1.5-3) COTO (2018b)

Category	Oxygen Permeability Index (OPI – log scale)	Water Sorptivity Index (WSI – mm/hr$^{0.5}$)	Chloride Conductivity Index (CCI – mS/cm)
Laboratory	>10.0	< 6	< 0.75
Full acceptance	> 9.4	< 9	<1.0
Conditional acceptance	9.0-9.4	9-12	1.0-1.5
Remedial acceptance	8.75-9.0	12-15	1.5-2.5
Rejection	< 8.75	> 15	> 2.5

As previously stated in Chapter 2, the stripping of formwork linked to strength development, which is a bulk property, is not a sufficient indicator of concrete quality near the cover layer, which is dependent on the curing, a near surface property. Furthermore, durability parameter test methods (OPI and WSI) have proven to be far more sensitive than compressive strength to detect changes in manufactured quality (compaction & mix design) and curing (Surana, Pillai, & Santhanam, 2017). Even though compressive strength (f_c) as a "performance" defining parameter for assessing durability has received great criticism in the literature, the reporting of f_c values is important for quality control purposes and for this reason, this parameter is also included as a separate material test in Section 3.1.3 (General Breakdown into Groups).

However, it should be reiterated that there is no necessary correlation between strength and durability, as indicated in *Figure 2-12* and *Figure 2-13*, due to the range of DI values from good to bad (OPI = 1.9 log scale & WSI = 11.3 mm/hr$^{0.5}$) which represent the quantifiable concrete durability parameters that are achieved in one of the projects analysed in this study irrespective of the strength grade of concrete. The variable nature of all specimens is obtained even though they are of the same strength grade of concrete (30 MPa) and standard laboratory cubes. The correlation between the coefficient of permeability using actual data and the diffusion or carbonation coefficient is discussed in Section 5.2.2 (Correlation between Permeability and Carbonation) which depends on the measured OPI, relative humidity and other empirical constants.

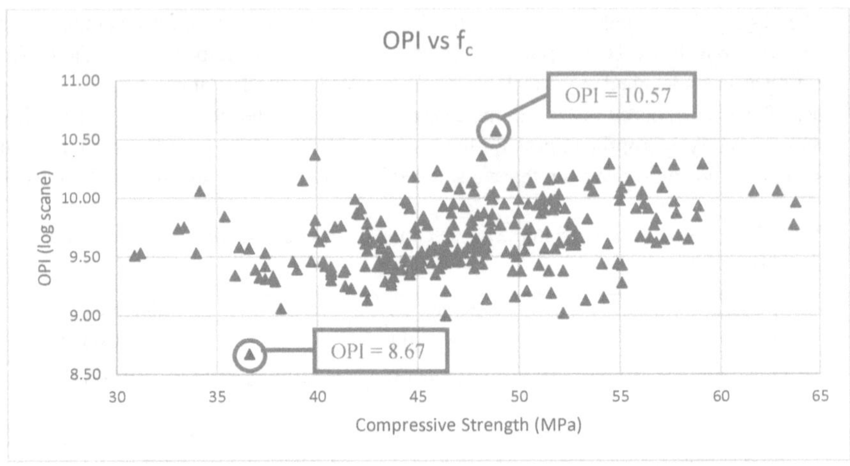

Figure 2-12 Relationship between OPI and f$_c$ (Project 5) (Source: Author)

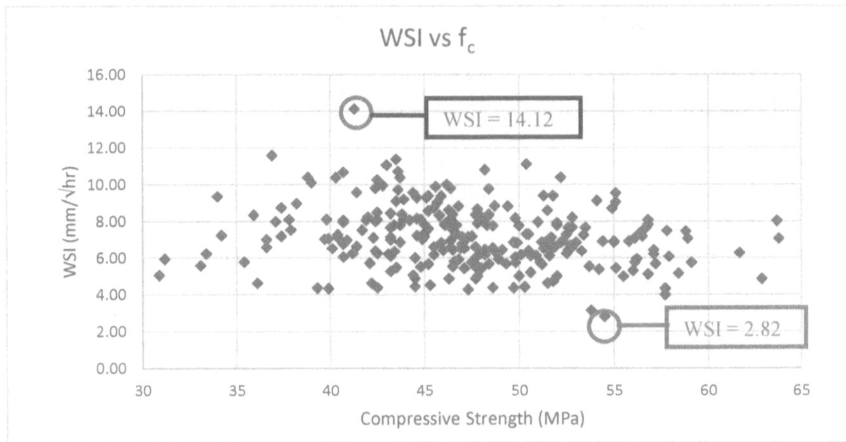

Figure 2-13 Relationship between WSI and f$_c$ (Project 5) (Source: Author)

Surana, Pillai, & Santhanam (2017) recommended that OPI is an applicable method to develop guidelines or performance specifications for the selection of curing compounds due to the subtle differences in values for different curing conditions. This is also evident since in *Figure 2.12* and *Figure 2.13*, the range in values for WSI is nearly a 10th in order compared to OPI for the same curing conditions. At first inspection, it seems as if WSI = 2.82 is an outlier, however upon further examination, the four determinations equal to 3.12; 3.14; 2.48 and 2.54, which equate to an acceptable CoV of 12.78 %. Another controversial point is that the strength achieved is on the high side of 54.5 MPa for standard 30 MPa concrete. Therefore, it would be better for curing efficiency to be measured in accordance with OPI which should be determined under trial conditions in the laboratory to assess the permissible deviations found in *Table 2-4*.

Although *Table 2-4* contained in COTO (2018b) refers to the different categories for acceptance or rejection, no reference to the relevant environmental classes, concrete cover or relevant binder in the case of CCI is made, which can create confusion on the procedure to be followed when analysing results. Although *Table 2-4* is used assess and classify project data in Chapter 5, it must be understood that results need to be assessed in line with their relevant environmental class and verification of concrete cover depth.

Therefore, ideally, *Table 2-4* should be expanded to form *Table 2-6* which provides more clarity to the adjudication of concrete quality. The environmental classes suitable for the general South African environments are listed in *Table 2-8*. Chloride index testing is required where specified by the engineer. Chloride conductivity testing should be used for materials selection and design of mixes in aggressive chloride conditions. It will therefore only be used as a check on mix designs during the initial stages of construction (based on trial panels) and on the test panels constructed. Testing should be undertaken during the construction and where a cement blend is used which is different from that given in the project specification, the appropriate limits should be discussed with the engineer for approval.

Durable concrete for environments where corrosion induced by carbonation presents the governing risk should conform with the desirable properties in *Table 2-9*. Durable concrete for environments where corrosion induced by chlorides in water or in the atmosphere presents the governing risk should conform with the desirable properties in *Table 2-10*. For durability design, the environmental class should be selected according to the structural element under consideration in *Table 2-5*, following which, the specified OPI or CCI limits can be read from *Table 2-9* and *Table 2-10*, respectively, for a maximum of three different cover conditions. Cover should also comply with the requirements given in *Table 2-7*.

It is necessary to split up the environmental classes for majority of the inland bridge substructures and superstructures, since they experience different types of exposure regarding carbonation-induced corrosion, such as XC3 and XC4, respectively. It is for this reason that the classification of the type of structure is found in Module 4 (Environment) and assessed dependent on the achieved DI values and cover depth found in Module 3 (Execution) further discussed in Chapter 4. The Durability Index Database (DIDb) must be able to detect when a Durability Index (DI) value is deficient in regard to *Table 2-9* and *Table 2-10*. These tables are contained in COTO (2018a) and are specific to RC structures which apply to a specific design and structure only – not a general case.

Table 2-5 Environmental Classes of Exposure for Elements of Structure (SANRAL Works Contract Proforma, 2019)

Element	Carbonation Environment (OPI)	Chloride Environment (Chloride Conductivity)
Foundations	n/a	XS1
Substructures	XC3	XS1
Superstructures	XC3	XS1

Table 2-6 Durability Parameters Acceptance Ranges 40mm cover (SANRAL Works Contract Proforma, 2019)

Acceptance Category	Test No./ Description/ Unit			
	Water Sorptivity Index (WSI – mm/hr$^{0.5}$)	Oxygen Permeability Index (OPI – log scale)		Chloride Conductivity Index (CCI – mS/cm) (Fly ash 30%)
		Substructures	Superstructures	
Concrete made, cured and tested in the Laboratory using Trial Panels	<6.00	>10.00	>10.00	<0.75
Full acceptance of in-situ using Test Panels	<9.00	>9.65	>9.85	<1.20
Conditional acceptance of in situ concrete based on results of Test Panels	10.50 – 11.50	9.25 – 9.40	9.45 – 9.60	1.40 – 1.60
Remedial acceptance of in situ concrete based on results of Test Panels	11.50 – 12.50	9.10 - 9.25	9.30 – 9.45	1.60 – 1.80
Rejection based on results of Test Panels	>12.50	<9.10	<9.30	>1.80

Table 2-7 Durability Parameters Acceptance Ranges: Cover for All Concrete Types (SANRAL Works Contract Proforma, 2019)

Test No.	Description of Test	Specified Cover (mm)	Acceptance Range	
			Min	Max
			Overall cover	Overall cover
B8106(g) (iv)	Concrete cover to reinforcement (mm)	30 to 80	85% of specified cover – 5mm	Specified cover + 15mm or where the member depth is less than 300mm the limit accepted in writing by Design Engineer

Table 2-8 Environmental Classes (Table A13.4.7-2) COTO (2018a)

Environmental class	Limited description
X0	No corrosion risk
Corrosion induced by carbonation	
XC1a	External concrete exposed to low humidity (<50% RH) and sheltered from moisture; arid areas; interior concrete.
XC1b	Permanently wet or saturated-damp.
XC2	Wet, rarely dry.
XC3	External concrete exposed to moderate humidity (50-85% RH) and sheltered from rain in non-arid areas.
XC4	External concrete exposed to rain or condensation, or alternately wet and dry conditions.
Corrosion induced by seawater, sea spray and saline groundwater	
XS1	Exposed to airborne salt but not in direct contact with seawater or inland saline water.
XS2a	Permanently submerged in sea (or saline) water.
XS2b	XS2a with abrasion.
XS3a	Tidal, splash and spray zones.
XS3b	XS3a with abrasion.

Table 2-9 Nominal Durability Index and cover values for 100-year service life in typical carbonating environments (Table A13.4.7-3) COTO (2018b)

Environmental class	Cover (mm), as specified	OPI (log scale)
For 100 year service life		
XC1a, and XC1b	40	9.15
	50	9.00
	60	9.00
XC2	40	9.40
	50	9.10
	60	9.00
XC3	40	9.65
	50	9.35
	60	9.05
XC4	40	9.85
	50	9.55
	60	9.30

Table 2-10 Nominal Durability Index and cover values for 100-year service life in typical chloride environments (Table A13.4.7-3) COTO (2018a)

Environmental class	Cover (mm), as specified	Chloride Conductivity (mS/cm)			
		Typical Cementitious Binder System			
		Fly ash (30 %)	Blastfurnace slag (50 %)	Corex slag (50 %)	Silica fume (10 %)
For 100 year service life					
XS1	40	1.20	1.30	1.60^2	n/a[1]
	50	1.85^2	1.95^2	2.20^2	0.40
	60	2.15^2	2.35^2	2.75^2	0.65
XS2a	50	0.85	1.00	1.20	n/a[1]
	60	1.25	1.45^2	1.70^2	n/a[1]
XS2b	60	1.10	1.30	1.55^2	n/a[1]
XS3a	50	0.65	0.80	0.95	n/a[1]
	60	0.95	1.10	1.40	n/a[1]
XS3b	60	0.85	1.00	1.30	n/a[1]

Notes: 1   n/a means cementitious binder system is not suitable for the indicated purpose
*   2   Maximum water: cementitious binder ratio for all binder systems shall be maximum 0.550*

2.4.2 Variability of Durability Index (DI) Results

The values in *Table 2-11* and *Table 2-12* are used to calculate the acceptance and rejection limits for single limit specifications, respectively. The acceptance limit (La) is the limit value of the sample mean within which a lot is accepted (COTO, 2018b). The factor (k_a) is used for determining the acceptance limits for single-limit specifications; the factor (k_r) is used for determining the rejection limits for single-limit specifications. For lower and upper limit specifications, the acceptance limit is defined as La and L'a, respectively. Conditional acceptance which is defined as acceptance of a lot at reduced payment in lieu of rejection will apply should the actual values exceed the above limit (COTO, 2018b).

Table 2-11 k_a values for assessment of Concrete Durability (Table A20.1.7-7) COTO (2018b)

Specified concrete durability property	Unit	k_a	La or L'a
Oxygen Permeability Index (OPI)	Log scale	0.25	Specified OPI − 0.25
Chloride Conductivity Index (CCI)	Milli Siemens/cm	0.20	Specified CCI + 0.20
Water Sorptivity Index (WSI)	$Mm/(hour^{0.5})$	1.50	Specified WSI + 1.50

Table 2-12 k_r values for assessment of Concrete Durability (Table A20.1.7-14) COTO (2018b)

Specified concrete durability property or cover (mm)	Unit	k_r	L_r or L'_r
Oxygen Permeability Index (OPI)	Log scale	0.40	Specified OPI – 0.40
Chloride Conductivity Index (CCI)	Milli Siemens/cm	0.40	Specified CCI + 0.40
Water Sorptivity Index (WSI)	Mm/(hour$^{0.5}$)	2.50	Specified WSI + 2.50

The rejection limit (Lr), on the other hand, is the limit value of the sample mean outside which conditional acceptance cannot be considered (COTO, 2018b). For lower and upper limit specifications, the rejection limit is defined as Lr and L'r, respectively. From *Table 2-11* and *Table 2-12*, the tolerances roughly correspond to a 1:10 chance of failure which is indicative of a 90% confidence level and correspond to margins of 0.3 log scale below for OPI, 0.2 mS/cm above for CCI and 1.0 mm/hr$^{0.5}$ above for WSI (Alexander, Ballim, & Stanish, 2008). Durability is a serviceability limit state criterion and hence the characteristic value can be set where 10 % (1:10) of the total area under the curve falls, therefore the characteristic durability can be defined as the durability of concrete below which not more than 10 % of the results are expected to fail. However, it should be noted that COTO (2018b) works with a mean value and not a characteristic value. In terms of concrete cover, the mean cover determined for each cover survey should exceed the specified cover minus C_{max} to avoid rejection of the lot according to *Table 2-13*. The lot complies with the requirements specified for concrete cover if the mean cover for the lot is not less than the specified cover minus the C_{ave} tolerance specified in *Table 2-13*, in which full acceptance will apply.

Table 2-13 k_a values for assessment of Cover (Table A20.1.7-6) COTO (2018b)

Specified cover (mm)	C_{max} (mm)	C_{ave} (mm)
75	15	5
60	12	5
50	10	5
40	8	5
35	7	5
30	6	5

2.4.3 Defining Outliers

An outlier occurs where one or more test determinations (in a sample) or results (in a lot) differ significantly from the remaining values obtained which could be ascribed to a random event or an assignable cause, respectively. In the case of the latter, this warrants further investigations, however for outliers defined within a sample, the differences can be attributed to a random event, in which the result should be discarded.

Outliers within samples can be further investigated by following Method 1 in COTO (2018b) which is very similar to computing the Z-score for a dataset, which simply put, calculates the number of standard deviations from the mean a data point is. Since four determinations exist within each DI result, sample outliers can be easily evaluated by inspection by calculating the effect on the CoV due to the exclusion of the values in question. In the case of outliers defined within a lot, the difference can be associated with an assignable cause which warrants further investigation.

In COTO (2018b), distinctions are drawn between first submission and resubmission of a lot for approval on the basis of conditional acceptance or rejection. Two conditions permit resubmission, namely, where the lot has been reworked such that a proper attempt was made to improve the unacceptable properties which is not applicable in the case of DI results or cover depth. The second condition is if there are valid technical reasons such as values out of the specification limits.

However, since DI results can be obtained from testing panels or in-situ coring from the structure, the resubmission of a specimen as stated in COTO (2018b) requires more clarity i.e. how DI specimens can be resubmitted. If the DI results from test panels do not meet the specification, then the payment adjustment factor should be applied to the entire lot in question. Should the contractor disagree with the results from test panels, then the contractor shall undertake coring of the structure at his own costs to negate the above hypothesis at positions decided by the engineer, giving priority to areas of low or inadequate cover. The contractor should be wary that in-situ DI values have been proven to be of lower quality than the results obtained from test panels, however occasional reversals have taken place in labour intensive contracts, therefore it is possible that full payment be reinstated should the specification be met (Alexander et al., 2008;. Ronny, 2011).

It is also possible that after coring the structure, conditional acceptance will change to remedial acceptance, where the contractor will be responsible for reinstating the structure to its desired state, by surface treatments or other acceptable means. Conditional acceptance with test panels could also lead to rejection from coring the in-situ structure. Reversals in reduced payments for DI tests, based on in-situ cores, could also occur where the in-situ test specimens were obtained from a greater depth where curing influence is negligible or where the heat of hydration resulted in greater maturity or at a much later age (not 28 days as for the test panel samples). Therefore, the results from in-situ coring of the structure should always take precedence over test panel results.

In order to determine theoretically whether there is significant difference between the two sets of test values (i.e. from two sets of test panels, in-situ cores or one of each), the Fisher F-test should be conducted first, and if necessary, also the t-test according to COTO (2018b). Should significance difference occur, then the second set of test values shall be regarded as a first submission. If no significant difference occurs, the first and second sets of test values can be combined for purposes of assessment.

2.4.4 Lot and Sample Sizes

A lot is a sizeable portion of work or quantity of material that is assessed as a unit for the purposes of acceptance control and selected to represent material or work produced by essentially the same process and from essentially the same materials (COTO, 2018b). Lot sizes are to be determined by the engineer considering the size and type of structure, specific portion and total quantity of concrete placed in a day and hence could vary significantly.

Within the PCM (*Figure 2-14*), construction lots are generated based on the relevant Quality Assurance (QA) scheme which determines test schedules based on project information and assesses test results in line with design specifications. For roads, in the Pavement Construction Module (PCM), this computerised system statistically determines the number and location of samples based on schedules of quantities for the input of lot results which subsequently calculates conformity and variability parameters based on design specifications that link to "action". This enables the automatic synchronisation of site data into with a central Integrated Transportation Information System (ITIS) database to conduct seamless as-built reporting. For small structures, samples can be combined of the same grade from different structures, provided the same plant is used and concrete is cast in the same period. However, a smaller size may be ordered by the engineer if the properties exhibit abnormal local variation within a normal lot size, the rate of production is high, or the area has obvious deficient quality which is in line with in-situ coring.

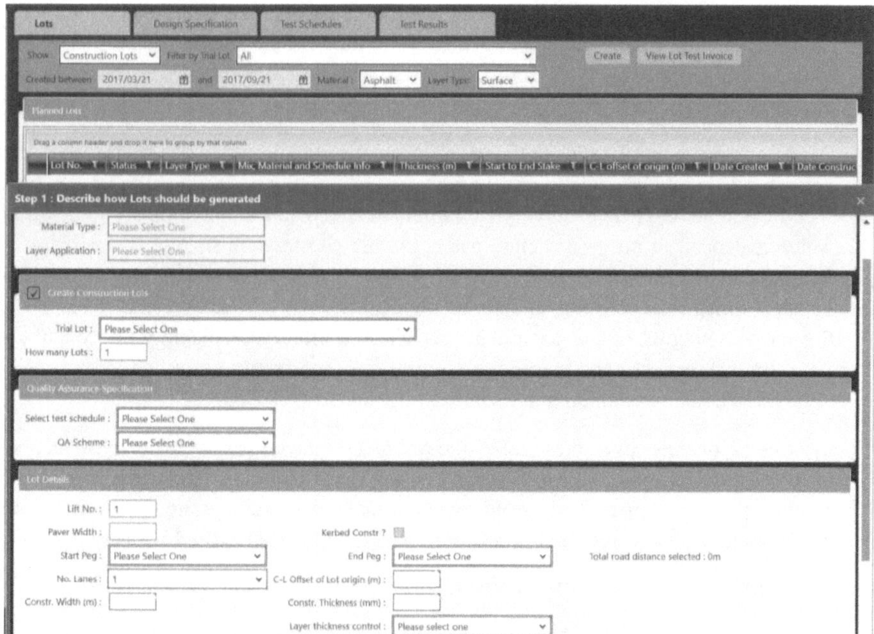

Figure 2-14 Determining Lot sizes in the Pavement Construction Module (PCM)
(SANRAL, 2017)

A random sample is a group of "n" test measurements at "n" separate test positions or on "n" sample portions obtained from the lot in an unbiased manner on condition that there are no other causes for rejection (COTO, 2018b). The lot size will be determined by engineer and therefore the larger the sample, the more reliable the result. Where test results deviate greatly from the remainder in the lot, they should be re-examined by further testing. If reasonable evidence to suggest the test was erroneous exists, it is regarded as an outlier and replaced with a fresh test result. If no such evidence exists and repeating or re-examining a test is impossible, which is commonly the case with DI results, since the material, manufacturing and testing conditions cannot be replicated, the area of the structure should be cored in-situ to determine actual performance for assessment with the specification. Therefore, when identifying outliers within lots (DI values out of specification), re-measuring any results that may possibly be defective should ideally occur with replacement of a specimen (test panel) or with coring the in-situ structure, for resubmission to be valid.

It is specified in COTO (2018b) that four complete tests will be conducted during the design phase which represents the trial stage DI results under laboratory conditions. However, during construction, the frequency of test panels subjected to DI requirements is minimised in certain instances. It is stated that during construction one complete set of tests are required for every $100m^3$ for the first $1000m^3$ and thereafter one set for every $500m^3$ of concrete cast (COTO, 2018b). It is possible that if less than $100m^3$ is cast in one day, then the provision for testing will fall away which reduces the lot sizes and hence the reliability of the results. Furthermore, $500m^3$ can constitute the entire volume of deck concrete for a medium scale bridge, whereby curing is of utmost importance and "one set" of complete tests would not accurately reflect construction quality occurring over 1 to 2 working days of full production

For acceptance control using statistical judgement principles, lot sizes need to be determined beforehand. In small, medium and large structures, minimum testing frequencies need to be specified in the onset. Typically, this information is known from preliminary to detailed design which is dependent on the structure geometry and construction stages. There is also no clear indication of lot sizes in COTO (2018b), besides the information which relate the concrete volumes which is a contributing factor to the inconsistency between projects. It should be noted that the grouping of various sample in lots will also affect the reliability i.e. concrete batched from the same sources and placed within the same time frames according to composition, but also with the same execution in the same environment.

In terms of concrete cover, the number of cover measurements per cover survey should be at least 12 over a nominal survey area of 600 mm x 600 mm and a minimum of 40 individual cover depth readings per square metre (m^2). A minimum of three cover surveys are required per lot and the minimum total area represented by a cover survey in a lot shall exceed 1 square metre (m^2).

2.4.5 Justifying a Maximum Variability or Percentage Defectives

Phi (Ø) is defined as the maximum percentage of a statistical population of values of a product property permitted to lie outside the specification limits where the product may still be regarded as being acceptable (COTO, 2018b). Acceptable values of phi (Ø) for DI results are not documented in specifications and is an aspect that needs attention when analysing these results for conformity. The relative measure of variability of a data set is best illustrated by determining the CoV within a project. The CoV is a relative measure of variability whilst the standard deviation is the absolute variability. When the CoV is greater than the repeatability standards for the relevant parameter being measured contained in Section 3.5.4 (Laboratory Equipment Used), then this is an indicator that there can be problems within the data.

When this parameter is sufficiently high, a large number of defectives can be present that result in exceeding the limit of 10 % according to the margins and confidence level stated in Section 2.4.2 Variability of Durability Index (DI) Results

Two terms have been frequently used to compare and differentiate between CoV measurements: these are, the within CoV (repeatability), which is the overall mean and standard deviation of the whole project, and between CoV (reproducibility), which compares the variability among projects. When mean values are substantially lower than the specification, the number of defectives increases even more considerably, as expected. In these instances, further investigation should be conducted to determine contractual penalties, remedial action and/or rejection. A payment reduction factor (fr) is a factor by which payment at contract rates is multiplied for calculating the payment for conditionally accepted work which shall be applied to the complete specific concrete member represented by the lot as indicated in *Table 2-14*.

Table 2-14 Fixed payment reduction factors for concrete (Table A20.1.7-13) COTO (2018b)

Property	Payment reduction
Concrete Cover	70%
Oxygen Permeability	80%
Chloride Conductivity	80%
Water Sorptivity	80%

In terms of concrete cover, the minimum cover depth achieved in a lot is defined as the characteristic value (percentile) below which a given percentage (typically 5 %) of all possible values of the cover depth population will fail. This percentile is defined as the value below which a given proportion of a collection of values such as a data sample or a whole population fails. For example, the 5th percentile of a population corresponds to the value below which 5 % of all theoretically possible values of the population will fail.

2.4.6 Monitoring Concrete Durability Performance Trends

In current specifications, such as COTO (2018b), schedules are provided that show the quantities and times for submitting materials for quality approval and mix design purposes, however for concrete, these are reflected as prescribed by the engineer. For quality approval only, this is handled between the contractor and concrete producer, which submits test results of the raw materials for approval. For mix design included, this must occur at least 4 weeks before structures items are affected on the construction program and will include laboratory testing.

Both quality approval and mix design information is important which assists in establishing the correlation from actual in-situ performance in terms of curing efficiency to long-term performance. Since lot sizes are determined by the engineer which depends on the size, type and specific portion of a structure as well as the total quantity of concrete placed in a day, lot sizes vary quite considerably on construction sites which has been detrimental toward standardising DI testing frequencies for RC structures with consistency. Even though probability density functions can be used and is presented in Chapter 5 to evaluate performance with respect to durability, due to the apparent advantages regarding the quantification of defectives, other measures can also supplement this to evaluate simple trends in DI results with time.

Defects identified either at a material or construction level need to be addressed in a timeous manner and this problem can be overcome by plotting a DI Summary for different time periods (*Figure 2-15*) that adopts similar rational as a target mean (CUSUM) chart used for compressive strength. As opposed to deducting the target mean from each test result and summing these differences up to form a cumulative sum, DI results can simply be plotted as per their achieved values. This simplification means that as more results become available, the plot can be continuously developed to identify changes in trends such as a reduction in mean values. In addition, correlations between different parameters can also be easily identified such as a decrease in OPI which might correspond to an increase in WSI as depicted in *Figure 2-15*. Therefore, the suggestion of plotting achieved DI values relative to the threshold value, strongly provides support to the COTO approach (Judgement Plan A).

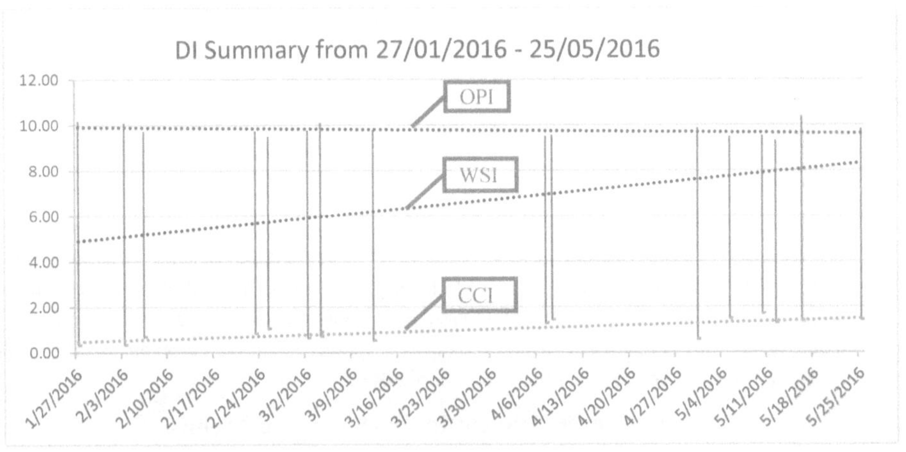

Figure 2-15 DI Summary for a filtered time period (Project 3) (Source: Author)

2.5 Summary

The literature review identified the two fundamental and elementary approaches to durability design, which are the prescriptive method and the performance or model-based method. It was further stated that durability design is an iterative process, with both methods being complementary in a complete design procedure, rather than opposite in nature which supports the use of hybrid approaches. Regarding both approaches for the durability of concrete in design codes and standards, the design work-flow procedure was provided. Factors that affect the durability of RC structures were found to be observational or experimental which supports the use of a hybrid database in order to contain both types of data.

Numerous studies were also found to only conduct experimental investigations to assess concrete durability under controlled conditions, even though the interplay of both experimental and observational characteristics during early-age site conditions has the greatest effect on long-term performance. From the discussion of the environmental exposure classifications, it was found that most international standards remained prescriptive in nature, despite the expansion of sub-classes in exposure conditions and alignment with their predicted severity of exposure during service life.

Examples of the classification of environmental action type and intensity as well as a service life prediction model which relates to performance-based specifications were discussed. In terms of prescriptive-based design approaches, it was stated that compressive strength which is a single characteristic parameter does not associate to the transport mechanisms affecting concrete durability, whereas the Durability Index (DI) values do. In terms of performance-based specifications, concrete durability has proven to be best defined, modelled and tested through its implementation, which has exposed limitations in the general rules governing prescriptive-based specifications.

A critique of durability provisions in relation to construction, specifically the quantification of concrete variability, quantification of concrete quality and quantification of curing effectiveness, was conducted. It was found that the sources of variation attributed to strength hold the same for durability parameters, which broadly stated arise due to the material, manufacturing and testing conditions. Regarding Durability Index (DI) values, it was evident that greater variability exists for in-situ vales by either assuming the same or lower averages for "as-built" quality in relation to material potential.

The sensitivity of the DI tests is a primary advantage over other prescriptive requirements that do not take into account material factors and construction effects. In order to quantify concrete quality, evidently the repeatability (single operator CoV) and reproducibility (between laboratory CoV) provide important information in order to specify limiting test values to obtain the required performance. Furthermore, semi-invasive testing has allowed for the development of performance-based durability design specifications in South Africa which consist of the coring of trial and test panels that are cured on site as a mechanism to ensure quality control for durability concrete.

However, it should be reiterated that the DI results from different mix designs and projects revealed that trial panels contained superior results than in-situ cores on 4 out of the 5 occasions considered for both OPI and WSI. If in-situ cores are not taken on site, the use of panels exposed to air and a wet-curing environment must be implemented to cater for this and also differentiate between the extremities of curing conditions. Ideally other aspects such as falsework and formwork as well as compaction and curing of samples should also be replicated during the above-mentioned procedure.

Lastly, a quality control scheme for concrete durability was proposed in relation to COTO (2018b). The goal of any sampling or judgement plan should be to distinguish good lots from bad lots where observations may be attribute or variable. Therefore, the kind of data to be analysed will determine the applicability of either an attribute sampling plan or variable sampling plan. Since DI results are measured on a numerical scale, the variable sampling plan is appropriate, however this is in contrast to the judgement plan proposed by COTO (2018b).

3   Database Design Methodology

3.1 Introduction

A database is a collection of information that is organised so data can be easily stored, managed, updated and retrieved. This chapter will focus on the design of the conceptual data model (DM) which establishes the basic concepts and the scope for the physical database. This process will establish the entities or data objects (distinct groups), their attributes (properties of distinct groups) and their relationship (dependency of association between groups)

Database design principles are needed to execute a good database design and essentially guide the entire process. Duplicate information or redundant data consumes unnecessary storage as well as increases the probability of errors and inconsistencies. Therefore, the subdivision of the data within the conceptual data model (DM) into distinct groups or topics in Chapter 3 which are broken down further into subject based tables in Chapter 4 will help eliminate redundant data.

The data contained within the database must be correct and complete. Incorrect or incomplete information will result in reports with mistakes and as such, any decisions made based on the data will be misinformed. Therefore, the database must support and ensure the accuracy and integrity of the information as well as accommodate data processing and reporting requirements. An explanation and critique of the current durability specification has been presented in Section 2.4.1 (COTO Durability Specification) since information is required on how to join information in the database tables created in Chapter 4 in order to create meaningful output in Chapter 5.

3.2   Database Design Principles and Lay-out

In this chapter, a conceptual data model (DM) will be designed such that DI results can be captured and structurally organised for further analysis based on the development of durability properties with change in material, manufacturing and testing conditions.

Hence, the key question of the research is:

"How can the influences of site practices (material, manufacturing and testing conditions) be measured or quantified such that inferences and correlations can be made to actual in-situ performance?"

The specific key questions for designing the conceptual and logical data model (DM) are:

1)   How should the data be subdivided into distinct groups or topics?

2)   What facts about each topic need to be identified and stored?

3)   What are the relations between the topics?

The first question will be answered as part of this chapter, in which the design of a modular structure or general layout for the physical database will be presented. The second and third questions will be answered in Chapter 4 and Chapter 5, respectively.

3.1.1 General Principles

The main objective of a database is to store, add, delete, update and manipulate data to make inferences. The output should also be presented in a logical manner. These objectives can be achieved by using database design principles as mentioned in the preceding section. Visser & Han (2003) raise the concern that if redundant data is present within the database, this may lead to errors. They further go on to summarise their general objectives which relate to four specific database design principles in total, such as to contain as little redundant data as possible, group only data of the same topic or subject define each set of data uniquely as well as define and maintain relations between grouped data.

3.1.2 Exclusivity of Data

It was previously outlined that in order to make inferences about the data to actual in-situ performance, the data (both DI values and cover depth) need to be captured in the correct places and be used together when assessing conformance with the specification. The database needs to contain such a system because the durability specification which although deterministic in nature, involves a rigorous approach to achieve concrete durability and therefore results in a "trade-off" between material quality and cover. For example, if the cover achieved is low (≤ 40 mm), more stringent criteria are applied on the OPI requirements and vice versa which ultimately affects the resulting payments. However, whilst the calculation of the relevant payment reduction factor is important and depends on the specification, the action in the event of non-conformity is also of relevance.

The exclusivity of data can be achieved by grouping data according to an exclusive form of identification which links to the same material, manufacturing and testing methods. This exclusive form of identification already occurs in the Bridge Management System (BMS) in which bridge structures (pre-fixed by B) and major culverts (pre-fixed by C) subject to concrete durability requirements are classified according to their Structure identification.

The advantage of having an exclusive form of identification eliminates unnecessary and redundant data. For instance, if the material variable (concrete constituents and proportioning) are stored as distinct groups for each DI result, repetition will result in unwanted and unnecessary redundancy which invalidates the second design principle according to Visser & Han (2003) that only data from the same topic should be grouped. Therefore, this information should only be entered once, for each project or mix design, which may or may not be different strength grades. However, a given mix design may change during a project due to the unavailability of materials such as coarse or fine aggregates, hence database entries should be able to be updated to cater for such scenarios.

Visser & Han (2003) state that a distinct group of data containing the material variable should be stored which satisfies the third design principle of exclusivity as previously outlined. Thereafter, only the unique identification codes (number or name) must be included in the material variable group, in which the fourth design principle is utilised to create the link, by defining relations between the two groups of data.

3.1.3 General Breakdown into Groups

The aim of the database is to store data from the compliance tests defined when developing the Bridge Construction Module (BCM) in Section 3.5.1 (Selecting the Test Methods). The Durability Index Database (DIDb) is essentially the BCM, which will be supplemented with other modules such as Geotechnical Investigation and Laboratory (aligned to the Pavement Construction Module).

Both OPI and CCI have been associated with various Service Life Models (SLMs), where the relevant DI parameter represents the as-built quality achieved after construction and serves as input to calculate the risk or probability of corrosion, among various other factors such as, environmental action or intensity, concrete constituents or proportioning and cover depth.

In order to store test results, one can create a group entitled "test results" and accumulate various results, however, such a strategy lacks purpose since it only lists the dependent variables in the analysis, that being the DI results and omits important independent variables which are the material, manufacturing and testing conditions, that occur during construction and are the variables that alter per construction site. These factors impact the as-built quality achieved after construction and must be defined with sufficient statistical ability to identify relationships between independent variables and dependent variables.

The use of performance-based specifications can result in durability predictions with a form of probability as assurance, whereas prescriptive-based specifications provides little or no indication as to the durability of concrete in relation to its transport mechanisms. Examples of the former approach is relevant to the general database design principles, however there is no form of probabilistic methodology used for durability at present in codes and standards in South Africa. Current examples include of such codes include the Probabilistic methods for durability design : DuraCrete (1999) (Netherlands), CSA A23.1/A23.2 (Canada), ACI/ASTM (United States of America), SIA262 (Switzerland) and CCES01-2004 Model Code (China).

In performance-based specifications, the material variable is dependent on many factors; however, the DI compliance tests have all been standardized, therefore differences will only result due to specimen specific conditions encountered on construction sites. These include the condition of the specimen itself, concrete composition, production/curing and exposure of the specimen. Thus, a breakdown of the specimen group can be further sub-divided into four distinct sub-groups, namely, Concrete Composition, Execution, Environment and Specimen.

Visser & Han (2003) state the significance that now since test results can be related closely to a specific specimen, that the source of data need also be part of the database as a separate group, namely "references". As previously stated in Chapter 1, the source of DI results in new construction can occur from testing panels or actual structural elements and be at different stages during a project life-cycle such as at mix design phase (trial panel) or during production (test panel or in-situ coring depending on specification). Therefore, the last group entitled "test results" will complete the 6-modular structure of the physical database as indicated in *Figure 3-1*.

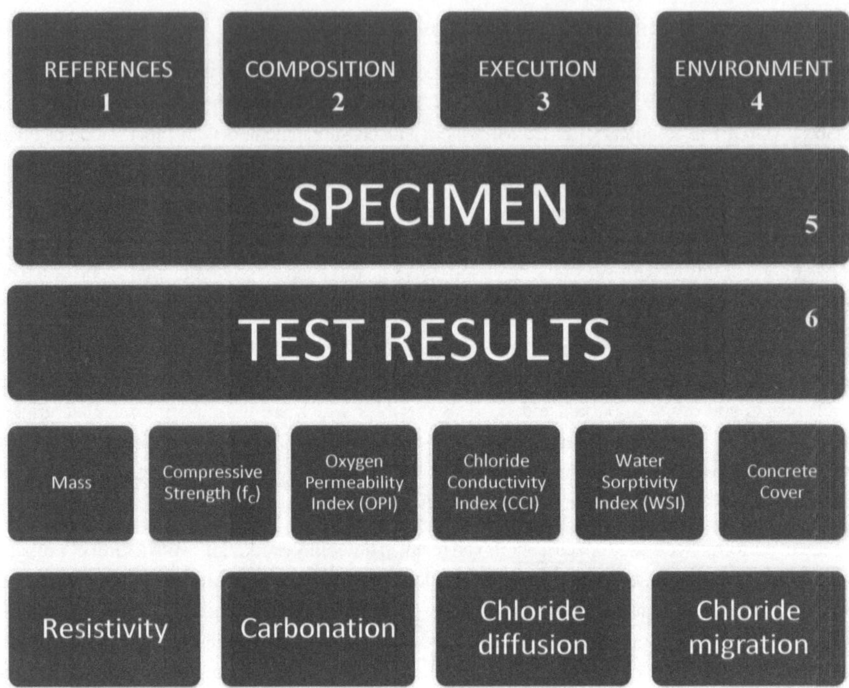

Figure 3-1 General layout of the Database Modules (Source: Author)

From the *Figure 3-1* it is evident that the specimen code is indeed the central code. In this way many significant database design advantages are achieved such as:

- Minimising information in the material test tables i.e. they only contain one unique identification code which is cross referenced between the test methods
- Several test methods can be performed on one specimen hence no duplicate information is stored by repeating the specimen code itself

Visser & Han (2003), mention that several specimens can also have the same composition and references as well as execution and environmental details, but choosing any other module as the central one would require more data storage and would therefore invalidate the first and foremost database design principle, to contain as little redundant data as possible. However, in the case of filling out the database for several specimens with the same information, besides material test results, special forms with consistent information need to be created for the database.

3.3 Database Preconditions and Requirements

3.3.1 Input Data

The aim of the database is the systematic collection and storage of concrete durability properties which enables future analysis for related research and development purposes. Examples of typical analysis, can include, inter alia, trend lines on DI parameters which associate to the material, manufacturing and testing conditions that can be applied judgment plans and quality control schemes as well as defining these DI parameters in terms of their associated distributions to be used in degradation models that form the basis from which a service life prediction can be made. Other types of statistical analysis can include correlation plots that measures the strength of association between two or more variables.

Visser & Han (2003), found that it was not necessary to include laboratory information such as the type of equipment used and the executor of the experiment. In the case of the Durability Index Database (DIDb), and, the DI tests which represent a form of performance-based specifications in infrastructure contracts, laboratory specific information is mandatory to report on repeatability and reproducibility standards further discussed in Section 3.5.4 (Laboratory Equipment Used) to ensure data reliability and no bias. Furthermore, contractual penalties arise from non-compliance hence by default, the above-mentioned details with other relevant project specific references is compulsory input for the database.

3.3.2 Output Data

The output of the database must evaluate the parameters or material tests enlisted in Section 3.5.1 (Selecting the Test Methods). Certain relationships between variables already exist, through years of research and should therefore guide the output and presentation of results to be incorporated into the physical database which is discussed in Chapter 5.

In some instances, there may be no relations established yet, or apparent when analysing data in isolation, therefore a standard output of the database should be set to form data tables for further processing by the user. Examples of these tables reported on by researchers are contained in various DuraCrete reports which underwent an initial process of checking and validation of the various relationships between the input and output data which could not be automated as that stage.

The database, should however, be able to filter appropriate data requested by the selection of various criteria specified by the user. Easy data manipulation provides a user-friendly platform to enable the creation of independent output. As another pre-condition, the selected filtered data should be able to be exported from the MySQL server to MS EXCEL (.xls) or MS WORD (.docx) format, as these are the most frequently used programs.

3.3.3 User Interface

A trivial requirement as mentioned by Visser & Han (2003) is that the database should be simple and hence user-friendly. This suggests that simplifications which enable easier data capturing should be investigated for the design of the conceptual data model (DM). One such simplification would include the use of the DI Spreadsheet Template (UCT, 2018a) in order to develop a user interface for the "material tests" group. Studies conducted by Nganga, Alexander, & Beushausen (2017) reveal that there have been problems in general with capturing information from sites. Some of the observations from the reporting of DI results in the past include missing information (results as well), the age of test samples and information on outliers. The most recent alteration to the DI Spreadsheet Template (UCT, 2018a) includes an additional worksheet for capturing information that can only be filled in from construction sites (Nganga et al., 2017). Therefore, one of the outcomes of this standardised form of reporting will be to characterise the independent variables according specific information regarding the material, manufacturing and testing methods.

3.3.4 General Use, Maintenance and Extensions

It is evident that the Durability Index Database (DIDb) or Bridge Construction Module (BCM) as part of the South African Road Design System (SARDS) for construction management and quality assurance will used by different parties. This would include those affiliated to SANRAL either directly or indirectly, such as project managers and consulting engineers or academics, for contractual obligations and limited to other R&D purposes. Therefore, another pre-condition of the database is that it can be run on common computer operating systems such as Microsoft Windows and Apple Mac OS X. The database design system chosen by Visser & Han (2003) involved Microsoft Access and therefore the application favoured Windows users, however this decision was taken 15 years ago, and as such, the world of information technology and computer science has rapidly evolved since then.

Moyana (2015) investigated, analysed and outlined the main limitations of a Microsoft Access system which ultimately prevents data capturing from different remote construction sites nationwide. Hence, the recommendations for the development of an open source relational database management system served on a MySQL platform is further investigated as the way forward for South Africa as depicted in *Figure 3-2* which will form part of the Integrated Transportation Information System (ITIS) that encompasses other important information such as the Bridge Management System (BMS).

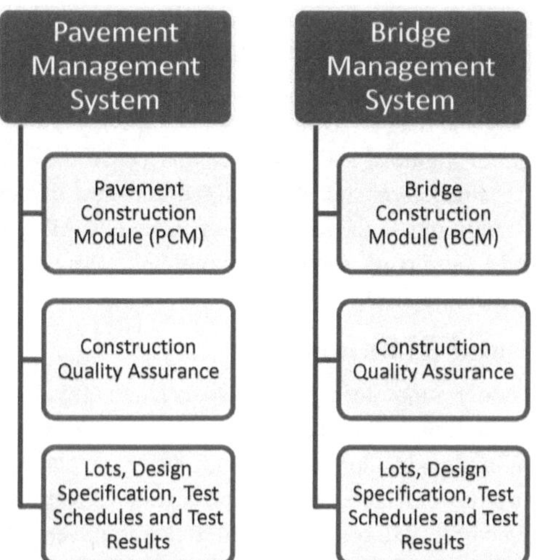

Figure 3-2 Overall Linkage Diagram – alignment of PCM and BCM for Asset Management

(Source: Author)

Once the system has been developed within ITIS, changes can only be made by the designers, including but not limited to the SANRAL Information Technology Staff. A separate platform will however be created for relevant users to select, filter, view and generate necessary output. Evidently, after experience from users, adaptations to the database can be made based on comments, however the initial structure of the database must be set up first to facilitate this process. Furthermore, it might be necessary to include other pertinent information, inter-alia, project specific field experiments or information on the "execution factor" as described by Visser & Han (2003), therefore a design requirement is that the database be formulated with a modular structure, whereby information or classes of data can be easily expanded without interference.

3.3.5 Exchange and Update of Data

In contrast to the working system in the Netherlands designed by Visser & Han (2003), the Durability Index Database (DIDb) or Bridge Construction Module (BCM) will be automatically updateable through the central MySQL server within ITIS and not similar to prior standalone programs developed by Microsoft Access. Visser & Han (2003) further go on to acknowledge that the standalone system does contain certain disadvantages, such as, the need for companies to send their database back to TNO (Netherlands Organisation for Applied Scientific Research) for manual updating procedures and the subsequent need to split the database up into two parts, each containing the data and program files, respectively. In order to prevent alterations to the program files by users, Visser & Han (2003) only distributed executable versions of the database, limiting any possible developer functionality.

Visser & Han (2003) also found it necessary to include a filter option to prevent the inclusion of "secret" data prior to sending the information back to the designer or maintainer. In the case of the DIDb or BCM, filter options to exclude certain data before submission is unnecessary for reasons of data integrity and data protection against tampering. However, while the project is in detailed design, structural engineers should take cognisance of the concrete durability targets and specifications to ascertain beforehand, the required measure of durability, in relation to the structure's exposure conditions in the specified environment. All test data, inter-alia, from trial panels, test panels and in-situ cores should be submitted, after which they will be analysed for variability, outliers and compliance.

3.4   Operations and Maintenance

The full-scale and long-term monitoring of the SA bridge network is indeed a costly affair, but the prioritisation of important structures will go a long way in rolling this out in cost effective stages. The successful implementation of a BMS for the Provincial Government of Western Cape (PGWC) outlined in a study conducted by Nell, Newmark, & Nordengen (2008) consisted of the inspection of 2300 structures (850 bridges and 1450 major culverts) from 2001 to 2003. In total, 175 structures were rated at a Priority Index (PI) < 60 and were red-flagged.

The defects primarily indicated general serviceability repairs and protections such as spalled or delaminated concrete and enhancement coatings, however the severity was much more pronounced in coastal/high rainfall areas. A background of the hydrochemistry of Southern African rainwater reveals that wet and dry cycles occur with total rainfall depths varying up to 100 % and more therefore the magnitude of rain water cannot be easily predicted (van Wyk, van Tonder, & Vermeulen, 2012).

Figure 3-3 Rainwater chloride concentrations (mg/l) in South Africa monitored between 2003 -2007 (van Wyk et al. 2011)

In addition, the study found elevated levels of both Na+ & Cl- due to high levels of windblown maritime aerosols in the coastal and immediate inland areas. The extremely high contents of Na+ & Cl- in rain or sea water contribute to surface moisture of concrete structures resulting in chloride penetration. This is mainly in the southern Cape's winter rainfall region and it must be stressed that rainwater Cl- concentrations are almost an order of magnitude larger than in summer rainfall regions as indicated in *Figure 3-3*.

In the study conducted by Nell, Newmark, & Nordengen (2008), major repairs were required by 20 of the structures; some of which included patch-repairs with coatings (deterioration), structural repair using bonded steel plates, external reinforced concrete elements and one complete replacement of a structure. It is clear that the repairs ranged from minor to extreme, however it is also clear that concrete structures can deteriorate in different ways which often do not result in one optimised repair method or cost-effective solution for all types.

It was also concluded that a number of additional repairs were required for the structures which were only identified at Detailed Assessment Report Stage. This was mainly attributed to the inability to make such judgements after the first visual inspection. Often detailed structural assessments, non-destructive and diagnostic testing was required in order to specify repair measures and hence a pre-repair Principle Inspection was indorsed in the BMS. In this sense, the linking of currently available DI values for structures nationwide to the BMS, can also advise to identify and prioritise structures in terms of anticipated maintenance and repair strategies.

3.4.1 Durability Index Database (DIDb)

There a numerous important factors in the success of database design. These are but not limited to the following. From the onset i.e. (conceptual DM stage), it is important to employ a data-driven approach and follow a structured methodology throughout the data modelling process. Therefore, data models (conceptual, logical and physical) should incorporate structural and integrity considerations as well as use normalization and transaction validation techniques in the methodology. Database developers also use diagrams to represent as much of the data models as possible and use a database design language. Another important process in database design is that of iteration, database developers work interactively with the users as much as possible and are willing to repeat steps as and when necessary. Users also play an essential role in the database design process confirming that the logical database design is meeting their requirements. Logical database design is made up of two steps and at the end of each step users are required to review the design and provide feedback to the designer. Once the logical database design has been 'signed off' by the users the designer can continue to the physical database design stage

DI compliance tests have been conducted on a majority of SANRAL construction sites since 2008, the inception of the performance-based durability approach based on pre-qualified concrete mix designs. The DI results vary by source (testing stage), project and structural element (bridge superstructure, bridge substructure or culvert), hence it is not practical to carry out a more comprehensive analysis by using the same project mix design parameters due to the tedious and expensive nature of the task (i.e. casting identical concrete mixes and curing under laboratory conditions).

Since the inception of DI performance-based specifications in 2008, over 12000 DI test results or determinations have accumulated within a repository at the University of Cape Town. However, manual processing and careful assembly of these test results that associate to a range of projects from across the country is not an easy task, often exacerbated by the lack of completeness within the data that prohibits accurate analysis. As result, the current process is deemed impractical and systems need to be investigated and designed to minimise the time taken for accurate data analysis. Concrete structures situated in severe cyclic wet and dry environments with a moderate humidity ranging from 50 – 80 % are at risk of carbonation-induced corrosion. Test results showing signs of lower permeability (OPI), higher sorptivity (WSI) or porosity values on average, need to be prioritised, as they are more prone to show signs of distress resulting in cracking and spalling. Furthermore, concrete structures found in extreme marine environments also need to be assessed for high chloride conductivity (CCI) or sorptivity or porosity as substructures and superstructures are at risk to chloride diffusion by saline seawater and airborne salt, respectively.

A complete database can perform this prioritisation for each failure mechanism (carbonation & chloride-induced corrosion) and condition of exposure (XC1a – XC4 & XS1 – XS3b) to determine the current "as-built" quality of our structures in relation to actual predictions in Service Life Models (SLM's) which can advise more accurately on maintenance and repair strategies. In addition, to confirm the applicability of locally used SLM's against actual conditions, the database of DI values for each exposure class can classify and inform on common trends in material, manufacturing or testing conditions in order to inform on later improvements for achieving concrete durability targets. Databases are commonly demarcated as either observational or experimental. However, the Durability Index Database (DIDb) will be one of the few exceptions, defined as a hybrid database containing test results from both observational and experimental conditions. The classification depends on the type of test result in question. Construction project or site data is obtained from observing the effect of process control and the environment (observational); laboratory/university research data depends on a set of closely controlled conditions or test methods (experimental). Both types of data can be useful, however each has its own limitations and advantages indicated in *Table 3-1* and *Table 3-2*, respectively.

3.4.2 Observational Databases

Examples of observational databases that exist currently have been assembled by SANRAL in their Pavement Management System (PMS) and Bridge Management System (BMS). Literature suggests that multi-collinearity occurs to some extent in all observational databases that can result in unstable parameter estimates which make it difficult to assess the effect of independent variables or predictor variables on dependent variables. In the case of existing structures, the predictor variables refer to visual condition assessment data obtained from inspections and the dependent variables refer to the calculated Overall Condition Index (OCI) based on the deduct method. Data-based multicollinearity results from a poorly designed experiment, reliance on purely observational data or the inability to manipulate the system in which data is collected. This type of multicollinearity is caused by a lack of balance between good and poor values but will be eliminated by including OCI and DI data at both target and non-target values.

Table 3-1 Observational Database Advantages vs. Disadvantages (Anderson, Luhr, & Antle, 1990)

No	Advantages	Disadvantages
1	Inexpensive to obtain or collect data	Multi-collinearity occurs where one predictor variable can be linearly predicted from others accurately
2	Minimal interference with construction processes	Prediction equations not useful outside region in which variables were observed hence MMT non-conformance is poorly characterised
3	If approximately same MMT used then performance can be predicted well	For the same MMT used the equations must be still be used cautiously to suggest methods to improve performance for projects constructed outside range of those observed

The advantages of observational databases well outweigh its disadvantages, since they are key in the model building and evaluation process since these databases defined execution and environmental parameters for degradation models that determine service life predictions (Visser & Han, 2003).

However, assuming the same MMT independent variables are used is a doubtful assumption in the case of RC structures which negates both the third advantage and disadvantage in *Table 3-1*. The variability of observational DI data creates difficulty in predicting performance by assuming the same MMT conditions occur, when this is commonly not the case.

Observational databases are successful for predicting performance in a PMS since greater quantities of works are executed to more measurable and consistent mechanical properties. However, in the case of RC structures, variations in the MMT independent variables, interrelated with other factors such as structure configuration (pre-cast, in-situ or composite), geometry (cross-section and span configuration) and method of construction (staged, involving temporary works etc.) implies that no two structures can be built identically within the physical database.

3.4.3 Experimental Databases

An example of experimental design consists of assessing test specimens exposed to air and a wet-curing environment on construction sites as this displays the extremities of curing conditions which affect the durability of RC structures (Surana, Pillai, & Santhanam, 2017). Experimental databases require a design dependent on the need for estimating certain coefficients (permeability, chloride diffusion etc.) in the response model which depends on the nature of the assumed effects. However, relating to concrete durability, these effects and interactions are nonlinear and hence experiments must be designed to cater and evaluate for this. The purpose of these kinds of databases is primarily to understand how the response of interest depends on the controllable factors (independent MMT variables) such that appropriate levels can be set.

Table 3-2 Experimental Database Advantages vs. Disadvantages (Anderson, Luhr, & Antle, 1990)

No	Advantages	Disadvantages
1	Range in variables can be specified to include all regions of interest (target values and non-conformance ranges)	Undoubtedly very expensive to develop
2	Importance of independent variables (MMT) and interaction on response of interest can be evaluated	Need for constructing poor performing specimens
3	Prediction equations valid over larger regions since independent variables (MMT) are controlled and varied in well-balanced manner	-
4	Prediction equations will suggest optimal set of independent variables (MMT) for purpose of developing acceptance strategy	-

Durable structures result in good DIs and in contrast, structures of lower quality will inevitably result in poor DIs, hence a fundamental problem arises in adequately assessing values that either lie in between or are outlying. This stresses the importance of properly defining a set of classes linked to different acceptance actions and boundary or threshold values linked to contractual penalties/remedial actions which is catered for in the DI performance-based specifications but relates to both experimental and observational conditions.

For the DIDb to be able to quantitatively relate non-conformance to serviceability, data must be contained at both target and non-target values. Therefore, experimental conditions must be designed to set DIs at both target and non-target values, and observational conditions must exist where DIs will serve as the monitoring parameters needed to ensure correlation to in-situ performance. The variability of experimental DI data is important to ensure non-conformance is not poorly characterised as per the second disadvantage in *Table 3-1*. This often results in the need to construct poor performing specimens which is also the second disadvantage in *Table 3-2*.

3.4.4 Hybrid Databases

TNO Building and Construction in the Netherlands created a prototype database for use in industry to collect durability data (Visser & Han, 2003). This is in line with other National Projects in France such as APPLET (Aït-Mokhtar et al., 2013) and PERFDUB (Linger & Cussigh, 2018). The main objective of the APPLET project was to quantify the variability of concrete durability properties to enable probabilistic performance-based specifications for service life prediction, whereas PERFDUB addressed setting up a methodology for performance-based specifications to justify durability of concrete in RC structures. However, a key driver for initiating the PERFDUB project was implemented by the French Association for Civil Works (AFGC) which set up a task group for the creation of a database dedicated to collecting concrete durability results (Carcasses et al., 2015).

Therefore, from the literature it is evident that examples of hybrid databases that currently exist consist of the TNO Building and Construction Research durability properties database for concrete design which contained experimental and observational data for both new and existing construction as designed by Visser & Han (2003) which is discussed in Section 3.4.4.1 (Database for durability properties of Concrete Design). Further application consists of the durability indicator database implemented by the French Association for Civil Works (AFGC) for data originating from laboratories as well as construction sites (Carcasses et al., 2015) which is discussed in Section 3.4.4.2 (Durability Indicator Database). Due to the relevance of the existing BMS in this study, it was found necessary to briefly discuss how Miyamoto & Nakamura (2003) developed the Japan-BMS to compute side-by-side service life predictions for existing RC structures according to durability and load-carrying capability which is discussed in Section 3.4.4.3 (Japan-BMS Bridge Rating Expert System). Finally, a proposal in the South African context is made for the DIDb based on the relevant literature which is discussed in Section 3.4.4.4 (Proposal for South Africa).

3.4.4.1 Database for durability properties of Concrete Design

Rijkswaterstaat is part of the Dutch Ministry of Infrastructure and the Environment in the Netherlands responsible for the design, construction, management and maintenance of diverse infrastructure facilities such as the national road and waterway network including an extensive flood prevention water system that is pivotal to the countries existence and protection. The coastline represents the most significant civil engineering accolade due to 13 series of dams and storm surge barriers which protects the country from the flooding of the North Sea. The sea often comes with arguably the most severe type of concrete degradation, chloride-induced corrosion and for a given amount of chlorides in the pore water; the corrosion risk is also higher for a carbonated concrete structure exposed to cyclic wetting and drying cycles. Hence the environmental exposure in this region is one of the most severe worldwide with a multitude of deterioration mechanisms consisting of freeze-thaw attack in addition to what is experienced under typical South African conditions.

Rijkswaterstaat in cooperation with five knowledge institutes in the Netherlands brought together their power in the formation of the Delft cluster which in 2003 published a report titled, "Database for durability properties of Concrete Design & Manual". TNO Building and Construction Research was instrumental in the process of setting up a prototype database for use in industry to collect durability data. Despite the definition of material and environmental variables in the DuraCrete project, a significant limitation was the relatively small amount of data and the fact that the defined variables displayed a stochastic feature. This means that the variables displayed a certain type of probabilistic distribution and hence one of the main findings from the project was that the uncertainty concerning the distribution type and corresponding parameters was relatively high.

The main objective of the database was to facilitate proper data collection and hence the reliability of estimates for service life based design of RC structures were increased (Visser & Han, 2003). A user manual for the completed prototype of this database was made available that contained input for information from both laboratory and field durability tests with quality control aspects. An important aspect to consider is that the database was set up as a modular system such that it could be easily expanded to comprise of new modules.

The filter tool was used to provide information on the ageing effect, influence of concrete mix design and manufacturing (environmental and execution) on the independent durability variables. Updating of the database was a critical pre-condition of the prototype, that involved at that stage, the manual collection of data tables from several different sources and updating it into a central database. In the next stage after evaluating the prototype, the update feature was scheduled to be designed once the general design of the database was approved.

3.4.4.2 Durability Indicator Database

The French Association for Civil Works (AFGC) have initiated the first phases of setting up a durability indicator database consisting of a working group of 10 members. The aim of this project is to improve the implementation of the performance-based approach in line with the objective of fib Model Code 2010: "to identify agreed durability related models and to prepare the framework for standardisation of performance-based approaches." There are also several levels of sophistication catered for including deemed to satisfy, partial factor and full probabilistic design approaches. One of the hurdles yet to be climbed by the French is simultaneously being faced in South Africa which needs to be verified. This is the anticipation of the variability of durability indicators all along a construction period to define a proper safety margin between characteristic and average values. Despite the inception of Table 6000/1: Concrete Durability Specification Targets, the significance of a given durability indicator threshold value is not obvious for the following reasons as stated by (Carcasses et al., 2015):

- Characteristic value is not easily determined by only a few values
- Minimal value risk (5 % or 10 %) is associated to the proposed value

The best way to take into account the variability of concrete and its standard deviation is to group durability indicators where certain independent variables are controlled. These will include the exposure condition, mix parameters (w/b ratio, binder type and content) and the test method. This will allow defining the distribution law of each durability indicator more precisely, know which law of probability should be used and hence quantify the variability. *Figure 3-4* indicates the flow chart of the global database used in the French context. The database aims at tentatively proposing control conformity rules to be used in the performance-based design approach.

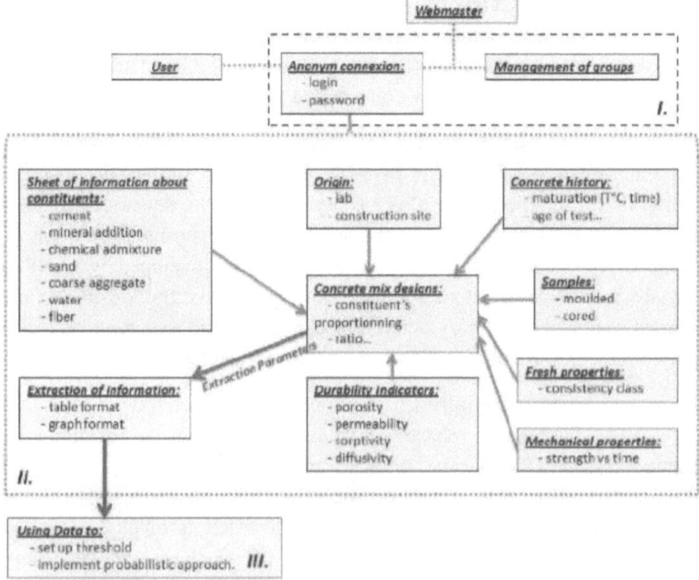

Figure 3-4 Structure view of database (Carcasses et al., 2015)

3.4.4.3 Japan-BMS Bridge Rating Expert System

In Japan, Miyamoto & Nakamura (2003) developed a computerised system that allows for two deterioration predictions for durability: the first occurs based on prevailing concrete deterioration processes (carbonation or chloride-induced corrosion); and the second computes predictions for load carrying capability deduced on visual condition assessment data found within the Japan-BMS. Arguably, both methods are pivotal in order to selecting repair/strengthening/maintenance plans on the basis of cost minimisation, quality maximisation and user safety. Miyamoto, Katsushima, & Asano (2013) further extended the Japan-BMS to include all Pre-stressed Concrete (PC) bridges which represent a large percentage of the structures found on South Africa's national routes. The Japan-BMS is also aligned with a PC modified Bridge Rating Expert System (BREX).

3.4.4.4 Proposal for South Africa

Moyana (2015) followed a systematic approach of transcribing DI data from construction sites and research experiments using two programmes, namely Microsoft Excel for spreadsheet generation and Microsoft Access for database management. One of the main limitations of this method is that all information from the database is saved into one file which cannot exceed 2 GB. The performance and response time of the database to concurrent users (> 20), possible corruption issues when using different operating systems and lack of web-based functionality in the reports generated are further limitations.

It was hence suggested that an open source relational database management system (MySQL) be hosted on a website with plugins to be adopted in future for the main advantage of automated data analysis and graph generation without much client user input and for ease of integrating new DI test rest results as and when they come available.

Therefore, the recommendations for the development of an open source relational database management system served on a MySQL platform is further discussed. ITIS (Integrated Transportation Information System) is a comprehensive tool developed by SANRAL in order to support management tasks and assist the technical decision-making process (SANRAL, 2017). The South African Road Design Software (SARDS) also developed by SANRAL is linked to the ITIS portal which to great extent provides a comprehensive framework to facilitate the process of pavement design. The primary objectives, conditions and requirements is that the Agency can share, interrelate and use information from different stakeholders.

SARDS system supports the infrastructure lifecycle and even though emphasis is placed on the Operation and Maintenance phase which relevant details are captured for the Pavement Management System (PMS), other phases such as planning leading to design and construction are also served with ITIS functionality. Therefore, since SARDS is linked to design investigations, performance simulations and construction quality assurance, it was found necessary to extend this software to RC structures for which relevant details are captured for the existing Bridge Management Systems (BMS) in ITIS.

The information contained in the BMS pertain to the location, environment, repair activities, maintenance schedules and structure condition (cracking etc.). These factors are related to the hybrid DIDb which although is primarily proposed for new construction, can also be easily extended to existing structures. There is significant value from combining the information from the existing observational database (BMS) to differentiate between structures and evaluate the importance of various factors affecting field performance.

At the end of construction, most MMT data is stored manually in massive amounts of files making retrieval and database assembly difficult for the completion of as-built records for projects. As-built records must be entered manually by consulting engineers through the ITIS portal, however with the implementation of SARDS BETA 2.2.10.11 developed by the South African National Road Agency Limited (SANRAL), information pertaining to the minimum sample size , testing frequency, classification of outliers, test data processing and laboratory equipment used for Quality Assurance (QA) schemes can be incorporated into the system to conduct seamless as-built reporting for test schedules and test results.

Including a QA scheme for RC structures in SARDS BETA 2.2.10.11 is profitable to serve as a platform to contain test results from a multitude of construction projects at the various stages (design and construction) which can be extended to operations and maintenance by its link with the BMS. This allows for the correlation of project specific information regarding material, manufacturing and testing conditions to take place in one central database.

Defects such as cracking can take many different forms and appear either months or years into the service life of the structure that compromise the cover concrete by providing ingress to harmful substances and contribute to reinforcement corrosion. However, on construction sites, from batching of the concrete, it is the actual transporting, placing, compacting and curing that ultimately determine the achievable extent of durability properties for new construction. Therefore, it is evident that defects are linked to three prevailing factors such as material, manufacturing (production and construction practices) and testing conditions, which affect primarily the microstructure development of the cover concrete.

Often, the only option is to rectify the damage when the problem arises, at the expense of the client. This is done through the BMS using information and ratings based on visual assessments to prioritise and highlight problematic structures based on their defects in a particular environment (exposure class) to calculate the Overall Condition Index (OCI) based on qualitative rating grades. By linking the abovementioned BMS to performance-based DI values, defects can be classified as durability or load-related, the two most common causes, with the statistical criteria required to make engineering judgements for both new construction and existing structures.

In this regard, the database will aim to highlight the differences in permeability and sorptivity values for laboratory and field cured concrete which can hence give an indication of the impact of external factors (material, manufacturing and testing conditions) under site conditions on concrete durability. The database can hence be used in the future to monitor the variability and interplay of observational and experimental conditions on concrete durability properties.

For structures with only a few test results (insufficient DIs), the main difficulty lies in assuring the results obtained can be used with a certain probability to conclude the in-situ (as-built) performance is above a certain threshold value. Hence, this is another issue that the database seeks to correct by creating a system to measure compliance. This system of compliance will link to the 'action' component when assessing test results for conformity with design specifications thereby informing on acceptance procedures, contractual penalties or remedial action.

Expert or "Knowledge-Based" systems deploy a collection of engineering judgement, rules of thumb, experience and intuition. Examples occur in the medical profession (diagnosis of bacterial infections), geology (location of valuable ore deposits) and computer system configuration. The primary difference is that this systems process knowledge whereas conventional programming processes data. Knowledge-Based systems are particularly applicable when knowledge available to predict performance is partially judgemental and subjective.

To classify defects as durability or load-related, engineering judgement is indefinitely required. Durability data whether experimental, observational or a combination of both requires a certain degree of engineering judgement based on the DI values and cover depth achieved in relation to the durability specification. Structure condition data is subjective (to a certain extent) since it is based on the level of expertise of the inspector. Therefore, in saying that, the applicability and integration of the two systems can prove very successful in building a platform for an Expert (Knowledge-Based) BMS in the future. Incorporating and linking DI results to structures found within the existing BMS will therefore facilitate and enable the best use of the database in the future that can be extended from new construction to existing structures.

These systems can be set up relatively simply through decision tress or in a more complex way, with the latter employing questions that are answered in terms of probability estimates (initiation of corrosion, structure OCI below limit, payment schedules etc.). If properly designed, the Knowledge-Based system can account for interaction between variables. Conventional and knowledge-based calculations have been proposed for the hybrid DIDb based on durability of new construction and specification limits contained in COTO (2018a; 2018b) for the main parameters affecting performance. Therefore, the interaction between variables such as DI values and cover depth have been investigated in this study. However, further applicability of such a Knowledge-Based system, integrated with the existing BMS would allow for conventional and knowledge-based calculations to occur based on load-carrying capability of existing structures.

3.5 Developing the Bridge Construction Module (BCM)

The development of a Bridge Construction Module (BCM) should as far as possible be aligned to the existing Pavement Construction Module (PCM). Therefore, the proposed QA scheme for the BCM is conceptualised from the PCM which represents an interactive quality assurance system created for the daily capture and analysis of test results within the South African Road Design Software (SARDS) system. Therefore, to develop the BCM, information is required regarding the test methods to be performed on durability (D-class) concrete, the sampling frequency for each test method, test data programming required, and the laboratory equipment used.

3.5.1 Selecting the Test Methods

The required test methods to be performed on durability (D-class) concrete is indicated in *Table 3-3*. The first test method deals with the preparation of test specimens which is the standard procedure for all sampling i.e. from panels (trial and test) as well as in-situ elements. The latter three test methods are reported from laboratory testing and measure the OPI, CCI and WSI, respectively. It should be noted that the last method has not been formalised through the SANS procedures although it is an important and frequently reported parameter.

Compulsory input for these test parameters are also contained within the DI Spreadsheet Template (UCT, 2018a) which is recommended to be used to develop a user interface for the input of the data as previously stated in Section 3.3.3 (User Interface). Input fields as contained in the referenced spreadsheet are consistent with the information contained in Module 6 (Test Results) further discussed in Chapter 4.

Table 3-3 Concrete Durability Index (DI) test methods (COTO, 2018b)

SANS Reference	Test Method / Requirement
SANS 3001-CO3-1	Concrete durability index testing - Preparation of test specimens
SANS 3001-CO3-2	Concrete durability index testing - Oxygen permeability test
SANS 3001-CO3-3	Concrete durability index testing - Chloride conductivity test
SANS 3001-CO3-4 (Proposed)	Concrete durability index testing - Water sorptivity test (UCT, 2018)

Other test parameters for fresh concrete (SANS 3001-CO1) and hardened concrete (SANS 3001-CO2/3) are defined below. For example, the latter consists of the compressive strength prepared in accordance with SANS 3001-CO2-2 and tested in accordance with SANS 3001-CO2-3. In addition to the latter, concrete cover readings obtained from electro-magnetic cover meter devices which do not fall under a specific SANS test method are also crucial to decide compliance with the durability specification. Therefore, these test methods are crucial and mandatory input parameters for concrete durability testing as initially decided in Section 3.1.3 (General Breakdown into Groups).

3.5.1.1 Fresh Concrete (SANS 3001-CO1)

For consistence or workability, the consistence class is measured depending on the contractor's chosen construction method. For self-compacting concrete (SCC) and pumped concrete there are additional requirements for viscosity, passing resistance, sieve segregation resistance and initial drying shrinkage capacity, however common structural concrete will comply with the below SANS standards or test methods as indicated in *Table 3-4*. It should be noted that these results are needed daily when concrete is batched or ready for casting on construction sites. Slump can vary from mix to mix, however in general, increasingly long waiting times or excessively high-water additions which is not permitted can be detrimental to the measured slump which depends on the specification of the concrete.

Table 3-4 Slump test methods (COTO, 2018b)

SANS Reference	Test Method / Requirement
SANS 3001-CO1-3	Slump test
SANS 3001-CO1-4	Slump < 10 mm and Vebe test is chosen measure
SANS 3001-CO1-5 (EN12350-4)	Degree of compactability is chosen measure
SANS 3001-CO1-6	Slump > 150 mm and flow diameter is chosen measure
SANS 3001-CO1-9 (EN 12350-8)	High workability concrete and slump flow diameter is chosen measure

For pumped concrete contained within most bridge decks, slump is determined as per SANS 3001-CO1-3 with a maximum upper limit of 175 mm. At mix design stage, the various tests include, inter alia, for bleeding / settlement (ASTM C232 or EN 480-4: where the later method is only applicable when the effect of the admixture dose is reported on), initial drying shrinkage capacity (SANS 3001-CO2-7) with a maximum of 0.04 % for Pre-stressed Concrete (PC) or 0.045 % for Reinforced Concrete (RC) or Mass Concrete (MC). The test methods required for coarse and fine aggregates are given in *Table 3-5* and *Table 3-6*, respectively.

Table 3-5 Coarse aggregates test methods (COTO, 2018b)

SANS Reference	Test Method / Requirement
SANS 3001-AG4	Flakiness index ≤ 35
SANS 3001-AG10	10 % FACT ≥ 150 kN (dry) or 110 kN (wet)
SANS 3001-AG10 (Project specific)	ACV ≤ 25 % by mass

SANS 3001-AG13 (Project specific)	Soundness of mudrock aggregate ≤ 15% mass loss

Table 3-6 Fine aggregates test methods (COTO, 2018b)

SANS Reference	Test Method / Requirement
SANS 3001-PR5	Fineness modulus ≤ ± 0.2 from approved
SANS 3001-AG5	Sand equivalent ≥ 65 %

The additional requirements in the case of Self-compacting concrete (SCC) are defined in *Table 3-7* and are project specific. For determining the viscosity and passing resistance of concrete, only one of the enlisted measures are required.

Table 3-7 Self-compacting concrete test methods (COTO, 2018b)

SANS Reference	Test Method / Requirement
SANS 3001-CO1-9 (EN 12350-8)	500 mm Flow time (Viscosity)
SANS 3001-CO1-10 (EN 12350-9)	V-tunnel Flow time (Viscosity)
SANS 3001-CO1-11 (EN 12350-10)	L-box ratio (Passing resistance)
SANS 3001-CO1-13 (EN 12350-12)	J-ring step (Passing resistance)
SANS 3001-CO1-12 (EN 12350-11)	Segregation portion (Sieve segregation resistance)

3.5.1.2 Hardened Concrete (SANS 3001-CO2/3)

The norm for quality control relies on compressive strength (f_c) testing of concrete cubes as well as the suite of DI tests (OPI, WSI & CCI). Compressive strength testing occurs most frequently i.e. consecutive or alternative trucks (every $6 - 12$ m^3) whereas DI testing is less frequent i.e. depending on production and project requirements as described in Section 2.4.4 (Lots and Sample Sizes). Therefore, it is possible to only have a few samples of concrete that contain results from either test method (both f_c and DI).

It should be reiterated that DI results can be obtained from various conditions, all relaying different yet at the same time important information. For example, at mix design stage, "trial" panels are cast under laboratory conditions (standard wet curing periods), whereas in the field, samples may be obtained from panels cast at the site ("test" panels) or from the actual structure itself (in-situ coring). As-built sheets are set out similarly in this fashion to enable consultants to fill in these results as well as those obtained from cover meter scans.

For as-built reporting, the issue has been how successfully to cross reference the cover depth values with f_c and DI results. According to fib (2006), as-built documentation for RC structures can be used to confirm the design assumptions or possibly give the basis for corrective measures via direct input of relevant parameters for the service life design. This information can further serve as the basis for condition control of the structure during its service life. The obstructing factor is that different methods are used for the cover meter scans (manual, grid or block survey methods) which contain outputs that are often extensive resulting in the difficulty to cross reference this information in the as-built sheets. To worsen the situation, the cover meter scans are typically only done at close-out or upon completion of the structure. Due to this, the following recommendation is made for the system (*Figure 3-5*).

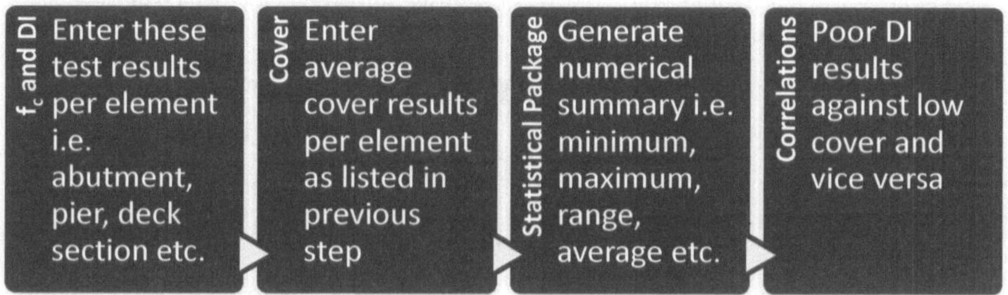

Figure 3-5 Test results system for hardened concrete (Source: Author)

3.5.2 Setting the Sampling Frequency

Since compressive strength has been used successfully as a performance defining specification for strength, the frequency of testing should be at the least similar for DI tests as performance defining specifications for durability. However, strength is a ULS criterion hence the characteristic value is set where 5 % of the total area falls under the curve (1:20 chance). Therefore, durability being an SLS criterion (1:10 chance) can be relaxed to a certain extent. For concrete strength, the lot size is dependent on the size and type of structure in which the concrete is placed including the specific portion of the structure as well as the total quantity of concrete placed in a day.

For this reason, it is stated that the lot sizes in concrete structures can vary considerably. It is stated that particularly in the case of small structures, it could be necessary to combine samples of the same grade of concrete from different structures, provided that the concrete is obtained from the same concrete plant and is cast in the same period using the same techniques. In the case of concrete durability, even if the same plant is used, either batched on site or from ready-mix suppliers, it is postulated that inevitable variability results from the concrete composition and the differences in execution and environment will result in increased variability which is a cause for concern if not computed.

For bridge piers or abutments that are constructed in two or more stages it would be necessary to have a test panel for each casting section per element. Typically, when the structure geometry and method of construction has been idealised, one can calculate the minimum number of test panels required. In the newly developed SARDS, each project is split up into its designated section of road length as indicated in *Figure 3-6*. An existing classification system for each structure already exists in the BMS which is discussed in Section 3.1.2 (Exclusivity of Data), therefore a simplification would be to cross reference the same information to the BCM using the Structure identification to split up the various bridges and major culverts for which data (DI values and cover depth) will be stored. Another aspect in common between the existing PCM and proposed BCM is the inclusion of results from the laboratory (*Figure 3-7*) at early stages in the project with those achieved in construction during full production, which must be maintained in the as-built reporting.

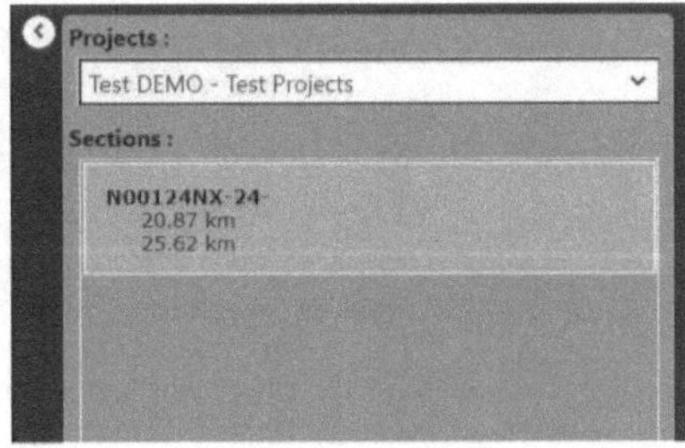

Figure 3-6 Projects selected according to National Route, Section and KM (SANRAL, 2017)

Figure 3-7 Different SARDS modules for Pavement Construction (SANRAL, 2017)

It is recommended that the BCM contain the minimum of six modules defined in Section 3.1.3 (General Breakdown into Groups). This system is pivotal for the database to function correctly as the data stored from test panels achieved during field conditions can be assessed in relation to what was achieved under standard wet curing conditions in the laboratory against the relevant acceptance and rejection limits defined in Section 2.4.3 (Defining Outliers). In some instances, with current project specifications, the DI requirement for laboratory results and field results were equal, which is incorrect. The specification must cater for the improved performance of laboratory results to understand and define a proper safety margin along a construction period.

In the case of bridges, or elements thereof, constructed utilising a half-width construction strategy, the sampling ratio should be increased to 3.3 percent of surface area of concrete placed i.e. 1 square metre (m^2) for every 30 square metres (m^2) surface area (COTO, 2018b). In addition, the entire area (rear and front) up to 1.5 m high above ground level on piers, walls and abutments should be fully tested before backfilling operations commence.

3.5.3 Test Data Programming Required

For projects with only one or two structures, the data processing need not be as extensive. Typically, the results can be split between substructure and superstructure since they are often different strength grades of concrete, cast during different construction stages and experience different types of exposure regarding carbonation-induced corrosion for majority of inland bridges. The database should however be able to filter results according to the six modules defined in Section 3.1.3 (General Breakdown into Groups) such that they can be plotted for different projects, types of structures and methods of construction indicated in *Table 3-10*. As such, the following search criteria are suggested and should thus be set as compulsory information for each structure. Evidently, the first links between the BMS and the BCM can be seen from the point of view that the search criteria proposed indeed have an effect on concrete durability. The necessity of the system to determine whether defects are structural or durability-related is beyond the scope of this study.

Table 3-10 Search criteria required for each structure (Source: Author)

Km	Type	Construction	Material
National Route km	Continuous with Expansion Joints Simply Supported Continuous without Expansion Joints (Integral) Composite	Precast Segmental Construction (PSC) Cast in Situ Balanced Cantilever Cable Stayed / Suspension Arch (Concrete or Steel)	Reinforced Concrete (RC) Pre-stressed Concrete (PC)

For projects with large amounts of structures, normality tests should be conducted which are primarily used to determine if a dataset is well-modelled by a normal distribution. Hence, the processing will include generating statistical summaries for the test data by conducting such normality tests, measuring the variability and creating distribution plots similar to work done by Moyana (2015) and Nganga (2011).

It should be noted that most common statistical tests rely on the normality of a sample or population which therefore stresses the importance of testing whether the underlying distribution is normal or at least symmetric. In general, the steps to be followed are to review the distribution graphically using histograms, box plots or QQ plots, analyse the skewness and kurtosis and employ statistical tests such as Chi-square, Kolmogorov-Smirnov, Lilliefors, Shapiro-Wilk (original or expanded), Jarque-Barre and D'Agostino-Pearson.

3.5.4 Laboratory Equipment Used

The SARDS modules are designed to track test results (DI values and cover depth) per piece of equipment and operator to detect any bias in data. In the case of the DI testing methods, the laboratory equipment used are stated in the relevant SANS standards with the allowed tolerances. This is a very important issue that links to the Repeatability and Reproducibility standards defined in Annexure A (Table A.1) of the oxygen permeability test and chloride conductivity test (SANS 3001-CO3-2 and SANS 3001-CO3-3) reproduced in *Table 3-11* and *Table 3-12*, respectively. By capturing the equipment and operator information, the precision and reliability of the test results can be associated to the below derived tables in which the validity of the data can be inferred, evidently linked to reliability.

Table 3-11 Guideline summary of Repeatability and Reproducibility values for OPI (SANS 3001-CO3-2:2015)

Repeatability and reproducibility	k-value	OPI
Repeatability	CoV (%)[1]	CoV (%)[1]
Laboratory data	$30.0 - 40.0$	$1.00 - 2.00$
Ready mix concrete data	–	$1.00 - 2.00$
Site data	$40.0 - 50.0$	$1.50 - 3.00$
Reproducibility	CoV (%)[2]	CoV (%)[2]
Laboratory data	$30.0 - 50.0$	$1.00 - 3.00$

Notes:　*1*　*Single operator coefficient of variation*
　　　2　*Between laboratory coefficient of variation*

Table 3-12 Guideline summary of Repeatability and Reproducibility values for CCI (SANS 3001-CO3-3:2015)

Repeatability and reproducibility	CCI
Repeatability	CoV (%)[1]
Laboratory data	$5.0 - 10.0$
Ready mix concrete data	$5.0 - 10.0$
Site data	$10.0 - 15.0$
Reproducibility	CoV (%)[2]
Laboratory data	21.1

Notes:　*1*　*Single operator coefficient of variation*
　　　2　*Between laboratory coefficient of variation*

3.6 Summary

The methodology was introduced in this chapter, in the form of database design principles which were aligned to the key questions of the research. In terms of the general principles for the database, it should be reiterated that the database should contain as little redundant data as possible, group only data of the same topic or subject, define each set of data uniquely, as well as define and maintain relations between grouped data. In terms of the exclusivity of data, an exclusive form of identification which links to the same material, manufacturing and testing methods will be implemented for grouping. The general breakdown into groups divided the data into 6 distinct groups, namely, references, composition, execution, environment, specimen and test results. Database preconditions and requirements were stated for the input data, output data, user interface, general use, maintenance and extensions as well as the exchange and updating of data.

Since the Durability Index Database (DIDb) will be used in the Design, Construction, Operation and Maintenance phases, it was necessary to discuss the importance of both observational data and experimental data for use in the hybrid database. Examples such as the Database for durability properties of Concrete Design in the Netherlands, the Durability Indicator Database in France and the Japan-BMS Bridge Rating Expert System are all examples of hybrid databases which support the current proposal for South Africa. Finally, in order to develop the Bridge Construction Module (BCM), alignment to the existing Pavement Construction Module (PCM) was required in terms of the test methods to be performed on durability (D-class) concrete, the sampling frequency for each test method, the test data programming required and the laboratory equipment used. The proposed QA scheme for the BCM was conceptualized from the PCM which represents an interactive quality assurance system created for the daily capture and analysis of test results within the South African Road Design Software (SARDS) system.

4 Results

4.1 Introduction

This chapter will focus on the design of the logical data model (DM) which adds extra information to the conceptual data model (DM) elements. This process will establish the database tables or basic information required for the database which represents the structure of all data elements, sets relationships between them and provides foundation to form the base for the physical database.

As mentioned in Section 3.3.5 (Durability Index Database), the logical database design phase of the methodology is divided into two main steps. In the first step, a data model (DM) is created to ensure minimal redundancy and capability for supporting user transactions. The output of this step is the creation of a logical data model (DM), which is a complete and accurate representation of the topics that are to be supported by the database.

In the second step, the Entity Relationship Diagram (ERD) is mapped to a set of tables. The structure of each table is checked using normalization. Normalization is an effective means of ensuring that the tables are structurally consistent, logical, with minimal redundancy. The tables were also checked to ensure that they are capable of supporting the required transactions and the required integrity constraints on the database were defined.

As mentioned in Section 3.1.1 (General Principles), the main objective of a database is the ability to store, add, delete, update and manipulate data to make inferences. One must also have substantial statistical ability to make these inferences, which requires that the database should be extensively filled out according to a QA scheme discussed in Chapter 3. The storage and manipulation of grouped data within the database is done using tables where each table must only contain information regarding the same topic.

As mentioned in Section 3.1.3 (General Breakdown into Groups), the database was defined as having 6 modules. Therefore, in this chapter, the tables used in each of the 6 modules will be further subdivided and clarification will be given on the chosen types of fields used in the tables. Once all the relevant information contained in the tables has been defined, then input forms with dedicated grouping of fields can be programmed in order to facilitate the filling out of the physical database. Specimen details relating to site and any additional information as required in the DI Spreadsheet Template (UCT, 2018a) also relate to input required for the different modules defined in this chapter.

4.2 Module 1: References

The references module will contain project specific information. The references table should hence be a quick reference to such project specific information, containing the following fields, as indicated in *Table 4-1*.

Table 4-1 References details

Field	Description
ID	Project reference identification
Project number	Project number
Project name	Project name
Consultant	Construction supervision
Contractor	Constructor
Laboratory	Laboratory

4.3 Module 2: Concrete Composition

Concrete's major constituents consist of cement or additions, which include supplementary cementitious materials (SCMs), admixtures, fine and coarse aggregates, such as sand and stone, and water. Various fresh concrete test methods have been outlined in Section 3.5.1 (Selecting the Test Methods) for the different aspects contained within this module which are included in the following tables.

It would be better for all of the abovementioned information to be summarised in one table to facilitate the speed of populating tables within the database. Therefore, predefined lists have been created in order to further facilitate this process. The condensed concrete composition details required are indicated in *Table 4-2*. However, in the case of aggregate, very limited choices are available, and users are encouraged to provide information on the aggregate type as per the relevant and approved mix design criteria.

Table 4-2 Concrete composition condensed details

Field	Description
ID	Concrete composition identification
Name	Short name
…Other fields	*Table 4-3* to *4-14*[1]
Comments	Other possible comments or specification criteria

Notes: 1   Tables refer to input via predefined lists or free-text

4.3.1 Cements and additions

Visser & Han (2003) state that, based on modern concrete design and data available in the literature, two different cements and additions should be allowed per composition since blends frequently occur in practice. Generally, all cement used during construction should comply with SANS 50197-1: Cement compositions, specifications and conformity criteria: Part 1 for common cements. In Appendix C, these options are listed according to their type and composition. The cement details are captured below in *Table 4-3*.

Table 4-3 Cement details for concrete composition

Field	Description
Cement type	See list in Appendix C
Cement strength	See list in Appendix C
Cement special	LH - low hydration / HS - highly sulfate resistant or LHHS - both
Cement content	Cement content (kg/m^3)

Visser & Han (2003) characterise additions by type and content only. Common concrete additions in South Africa are indicated in *Table 4-4* and should be selected accordingly and by specifying their content. The additions details are captured below in *Table 4-5*.

Table 4-4 Binder pre-defined list for input form

ID	Binder type	Binder Name
1	FA	Fly Ash
2	SF	Silica Fume
3	GGBS	Ground Granulated Blastfurnace Slag
4	GGCS	Ground Granulated Corex Slag
5	Other	Metakaolin, Calcinated Clay etc.

Table 4-5 Binder details for concrete composition

Field	Description
Binder type	See list in Table 4-4
Binder content	Binder content (kg/m^3)

4.3.2 Water and admixtures

The source and water content was found necessary to indicate according to *Table 4-6* and *Table 4-7*, respectively. More than one admixture are often required in a concrete mix and for this reason, the user can select the relevant admixture codes as in *Table 4-8* and specify their corresponding dosage or content in *Table 4-9*.

Table 4-6 Water source pre-defined list for input form

ID	Water source
1	River / lake
2	Municipal
3	Borehole
4	Other

Table 4-7 Water details for concrete composition

Field	Description
Water source	See list in *Table 4-6*
Water content	Water content (kg/m^3)

Table 4-8 Admixture pre-defined list for input form

ID	Admixture Code	Admixture Name
1	P / WRA	Plasticiser / Water Reducing Agent
2	SP	Super Plasticiser (High Range Water Reducing Agent)
4	AEA	Air Entrainment Agent
5	SetAcc	Set Accelerator
6	HardAcc	Hardening Accelerator
7	SetRet	Set Retarder
8	WrestA	Water Resisting Agent (Water proofer)

Table 4-9 Admixture details for concrete composition

Field	Description
Admixture types	See list in *Table 4-8*
Admixture content	Admixture content (kg/m^3)

4.3.3 Coarse and Fine Aggregates

The required list for users to select aggregates from is contained in *Table 4-10* with other coarse aggregate information to be included in *Table 4-11*.

Table 4-10 Coarse aggregate pre-defined list for input form

ID	Prefix	Aggregate / Type Name
1		Andesites
2		Dolerites
3		Dolomites / Limestone
4		Felsites
5		Granites
6	Transvaal, Cape etc.[1]	Greywacke
7		Quatzites
8		Tillites
9		Siltstone
10		Sandstone
11		Other

Notes: 1   Aggregate types (rocks) vary greatly depending on geological source/ locality

Table 4-11 Coarse aggregate details for input form

Field	Description
Coarse aggregate types	See list in *Table 4-10*
Content	Aggregate content (kg/m^3)

Additional information to fine aggregate type based on its source and amount is also indicated in *Table 4-12*. Users may select options from the predefined list in *Table 4-13*.

Table 4-12 Fine aggregate details for input form

Field	Description
Fine aggregate types	See list in *Table 4.13*
Content	Aggregate content (kg/m^3)

Table 4-13 Fine aggregate pre-defined list for input form

ID	Prefix	Aggregate / Type Name
1		River sand
2		Pit / Quarry sand
3		Dune sand
4	Granite, Cape Flats etc.[1]	Crushed sand
5		Other

Notes: 1 Fine aggregate types vary greatly depending on type and source/ locality

4.3.4 Fresh Concrete

The number of fresh concrete properties to be recorded can be very extensive. It is therefore better to limit these options to those indicated in *Table 4-14*.

Table 4-14 Fresh concrete properties

Field	Description
Temperature	Immediately after batching (°C)
Air volume	Air content in fresh concrete (%)
Vol. mass	Volumetric Mass / Wet Density (kg/m^3)
Workability	SANS 3001-CO1-3 (Maximum slump < 175 mm)[1] Vebe test (Slump < 10 mm) – SANS 3001-CO1-4[2] Degree of compactability - SANS 3001-CO1-5 (EN 12350-4)[2] Flow diameter (Slump > 150 mm)– SANS 3001-CO1-6[2] Slump flow diameter (high workability) – SANS 3001-CO1-9 (EN 12350-8)[2]
Viscosity	500 mm Flow time - SANS 3001-CO1-9 (EN 12350-8)[3] or V-tunnel Flow time - SANS 3001-CO1-10 (EN 12350-9)[3]
Resistance	L-box ratio - SANS 3001-CO1-11 (EN 12350-10)[3] or J-ring step - SANS 3001-CO1-13 (EN 12350-12)[3]
Sieve segregation resistance	Segregation portion - SANS 3001-CO1-12 (EN 12350-11)[3]

W:b ratio	Water - binder ratio

Notes: *1*   *Commonly used for most bridge decks with a maximum upper limit as indicated*
       2   *Chosen test method as per the project specifications and with slump characteristics as indicated*
       3   *These parameters are only required in project specific circumstances where SCC is prescribed*

4.4 Module 3: Execution

This module consists of three tables as shown in *Figure 4-1*, with the main table being the execution one, containing a further three classes of information, namely, curing, production and quality (both concrete and cover). Similarly, as for concrete composition, all information will be stored in one table, with one exception. Curing is separated into its own table since different execution regimes can have the same curing regime (Visser & Han, 2003). For instance, the separation of this table allows different execution regimes to be grouped or filtered according to production and curing type.

The fabrication or production type which commonly differs depending on the element considered is also contained within the same table. Correlational interconnections are made for both cover quality and concrete quality relating to the post-verified cover depth and measured DI parameter as defined in Section 4.6 (Module 6: Test Results). Evidently, both these parameters will have multiple results for the same element, in which the system must be able to cater for additional entries as and when necessary for Durability Index (DI) values and corresponding cover depth results. Visser & Han (2003) state that their first attempt, to have three separate tables for each class of information, was a failure as it invalidated some of their design principles, specifically the time taken to fill out information.

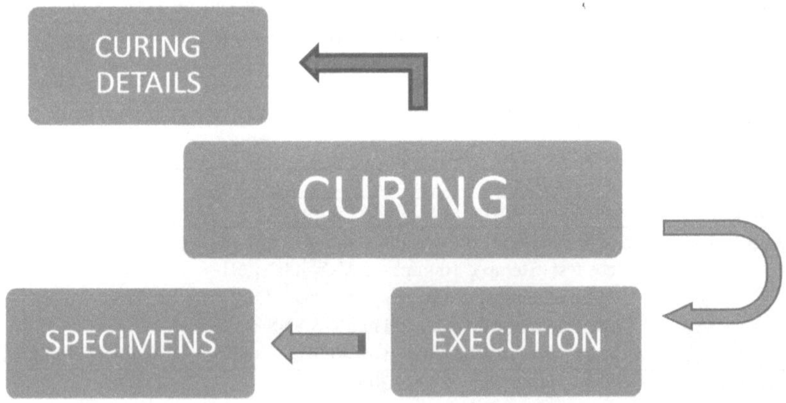

Figure 4-1 Arrangement of the tables within Module 3 (Execution)

4.4.1 Execution

The main table contains the identification number of the execution and its corresponding short name in *Table 4-15*. For curing, only a reference to the identification of the curing regime is stored that can be used to create a relation without storing additional information within this table. Furthermore, comments can be added concerning the execution procedure.

Table 4-15 Main execution details

Field	Description
ID	Execution identification number
Description	Short descriptive name for execution
Curing regime	Identification name for curing regime
Production-type	Continuous with Expansion Joints, Simply Supported, Continuous without Expansion Joints (Integral) or Composite
Construction-type	Precast Segmental Construction (PSC), Cast in Situ, Balanced Cantilever, Cable Stayed / Suspension or Arch
Material-type	Reinforced Concrete (RC), Pre-stressed Concrete (PC) or Structural Steel
Cover quality	Measured cover[1]
Concrete quality	Measured DI parameter[2]
Compaction	Compaction method used
Comments	Execution procedure comments - application of surface coatings or impermeable membranes (curing compounds)

Notes: 1   *Defined in terms of mean and standard deviation per element considered according to specification*
          2   *Acceptance categories for DI values (OPI, WSI and CCI) defined in Table 2.4*

4.4.2 Curing

Having its own separated table, curing is now defined by means of two interconnected tables, namely Curing regime (*Table 4-16*) and Curing details (*Table 4-17*). The curing regime is linked to the curing details by means of the same identification numbers. The former, similar to the main table, contains only the identification of the curing regime and its corresponding short name and for the latter, each period which contains a curing regime can be stored. Users are encouraged to select the curing type by means of a pre-defined list as in *Table 4-18*. Certain types of curing found in *Table 4-18* are associated to the temperature and relative humidity of the curing environment which are relevant to both the field conditioned test specimens and the conditions subjected to the actual structure.

Table 4-16 Curing regime

Field	Description
Curing regime ID	Identification number for curing regime record
Regime name	Identification name for curing regime

Table 4-17 Curing details

Field	Description
Curing detail ID	Identification number for curing detail record
Curing regime ID	Identification number for curing regime
Period	Period in curing regime that record belongs to
Unit	Unit of record (hours, days, weeks or months)
From	Start date of curing period
To	End date of curing period
Type	See list in *Table 4-18*
Temperature	Temperature (°C)[1]
Rel. humidity	Rel. humidity (%)[1]

Notes: *1*   *Additional parameters as required for curing type field number 4, 5, 7, and 8*

Table 4-18 Curing type pre-defined list for input form

Field	Curing Type	Curing Type Description
1	SUB-W	Submerged in water
2	SUB-L	Submerged in saturated lime water
3	FOG	Fog room
4	AIR	Indoors at constant T (°C) and RH (%)
5	AIR-C	Air with curing compound[1]
6	STEAM	Steam cured
7	Out / Shelt.	Outdoors – sheltered[1,2]
8	Out / Unshelt	Outside – unsheltered[1,2]
9	Cyclic	Cyclic (wet-dry)
10	Sealed	Sealed with polyethylene film
11	Covered	Covered with burlap

Notes: *1*   *Additional field to be entered is the curing compound type*
           2   *Average monthly values to be obtained from site weather station*

4.5 Module 4: Environment

Similarly, to Section 4.4 (Module 3: Execution), this module consists of three tables as shown in *Figure 4-2*, with the main table being that for the environment, containing a further three classes of information, namely, exposure, environment and structure details. As for curing, exposure is also kept in its a separate table. This is done since different environments can have the same exposure which can be split according to the different environmental classes contained in *Table 2-8* as well as to maintain consistency between the execution and environment modules. For instance, the separation of this table allows different environments to be grouped or filtered according to exposure types and structure details.

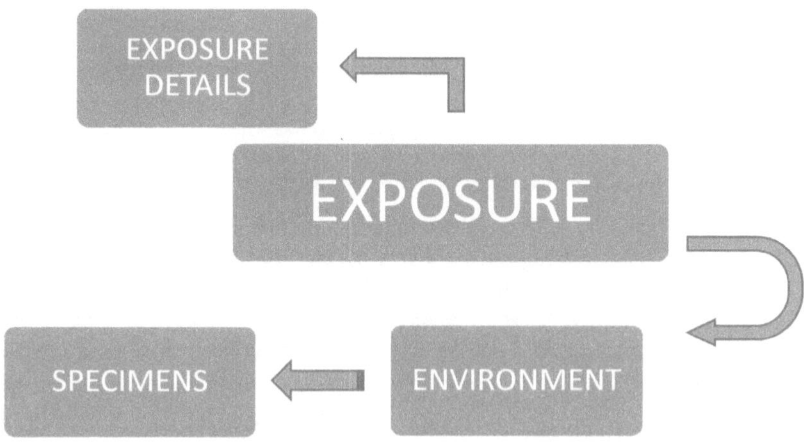

Figure 4-2 Arrangement of the tables within Module 4 (Environment)

4.5.1 Environment

The environment represents the locality of the structure and is concerned with the characteristics that are in common or influence the chosen exposure conditions. The main table contains the identification number and its corresponding short name for the environment as indicated in *Table 4-19*. For the different types of exposures, only a reference to the identification of the curing regime is stored that can be used to create a relation without storing additional information within this table. Furthermore, comments can be added concerning the environment.

Table 4-19 Main environment details

Field	Description
ID	Environment identification number
Description	Short descriptive name for environment
Exposure regime	Identification name for exposure regime
Structure type ID	See list in *Table 4.20*
Name	Name of the structure pre-fixed by B (bridge) or C (major culvert)
Location	Region and national or provincial route km reference
Orientation (structure)[3]	See list in *Table 4-21*
Main wind direction[3]	See list in *Table 4-22*
Avg. Temperature[3]	Avg. Temperature (°C)[1]
Avg, Rel. humidity[3]	Avg. Rel. humidity (%)[1,2]
Avg. no. of rain days[3]	Avg. no. of rain days[1]
Dist. coast[4]	Distance from the coast
Height sea level[4]	Height above the mean sea level
Dist. marine surface[4]	Distance from the marine surface
Distance water table	Distance from foundation level to mean water table level (+ = above, – = below)

Comments	Comments concerning the environment: dry, wet, windy, humid or cyclic wet / dry

Notes: *1*   *Average (mean) daily values to be based on 10-year recorded data*
    2   *RH data has displayed distribution characteristics like that of a normal distribution but should be evaluated against beta or Weibull (max) distribution types for applicability after fib (2006)*
    3   *Parameters required for carbonation-induced corrosion*
    4   *Parameters required for chloride-induced corrosion (SA Version of EN206 Standard descriptions)*

From the above table, the structure and its location within the environment is described in detail. To facilitate the filling out of the database and keep to original design principles, pre-defined lists have been created for the most critical information. Not all information listed in the above table will be required for each structure. For instance, when programming the input forms, only the relevant fields should pop-up according to the critical corrosion mechanism (carbonation or chloride).

In rare cases, distress mechanisms such as those listed above, or even others such as soft water attack, acid attack, sulphate attack and alkali aggregate reaction, can act in conjunction with others depending on the locality, exposure regimes and concrete composition. Therefore, it is important that site specific information such as the coarse aggregates used in concrete and external sources of aggressive agents are well documented.

Table 4-20 Structure type pre-defined list for input form

Field	Structure Type	Direction of proposed drilling
1	In-situ bridge decks	Vertical
2	Bridge piers or abutments	Horizontal
3	Precast beams (specify type)	Vertical Horizontal
4	Bridge / culvert parapets	Horizontal
5	Culvert walls / wing walls / slabs	Vertical Horizontal
6	Retaining walls	Horizontal
7	All bases (spread footings or piled foundations	Vertical
8	Piles	Vertical

Table 4-21 Structure orientation type pre-defined list for input form

Field	Orientation
1	Horizontal
2	Vertical
3	Inclined

Table 4-22 Main wind direction pre-defined list for input form

Field	Main Wind Directions

1	N
2	NE
3	E
4	SE
5	S
6	SW
7	W
8	NW

4.5.2 Exposure

Similarly, to curing and having its own separated table, exposure is now defined by means of two interconnected tables, namely Exposure regime (*Table 4-23*) and Exposure details (*Table 4-24*). The exposure regime is linked to the exposure details by means of the same identification numbers. The former, similarly to the main table, contains only the identification of the exposure regime and its corresponding short name with comments. For the latter, each period which contains an exposure regime can be stored.

Table 4-23 Exposure regime

Field	Description
Exposure regime Id	Identification number for exposure regime record
Regime name	Identification name for exposure regime
Comments	Comments concerning the exposure regime

Table 4-24 Exposure detail

Field	Description
Exposure detail Id	Identification number for exposure detail record
Exposure regime Id	Identification number for exposure regime
Period	Period in exposure regime that record belongs to
Unit	Unit of record (hours, days, weeks or months)
From	Start date of exposure period
To	End date of exposure period
Environmental class	No corrosion risk (X0), Carbonation (XC) or Chloride (XS)[1]
Aggressive agent	See list in *Table 4-25*
Concentration	Concentration
Concentration unit	Unit of the concentration

Notes: 1   Environmental classes for RC structures defined in Table 2-8

Table 4-25 Aggressive agent type pre-defined list for input form

Field	Aggressive Agent Type
1	Carbon dioxide[1]
2	Chloride ions[2]
3	Sulphate ions[2]

4	Seawater (multiple ions)[2]
6	Magnesium sulphate ions[2]
7	Acid[2,3]

Notes: *1   CO2 content ranges from 350 ppm to 380 ppm corresponding to concentrations of 0.00057 kg/m³ and*
0.00062 kg/m³ with an increase of 1.5 ppm per year and maximum standard deviation of 10 ppm (fib, 20C
2   Laboratory testing should be conducted to determine the concentrations, however where no such data
exists, assumed values can be taken from literature depending on the severity of the environment
3   Common acids encountered in groundwater involve sulphuric, hydrochloric and carbonic which can affe
concrete mix design depending on the source of water and underground or buried structures depending c
the mean water table fluctuations

4.6 Module 5: Specimens

This module serves as a reference point for the different types of samples (cubes, trial panels, test panels and cores extracted from the as-built structure) as well as to store important specimen details such as the dimensions and condition before testing in a single table. A record must be kept for all specimens received from site that details whether the specimens have been properly packed and if there is any damage on the cores assuming that they were extracted from panels (UCT, 2018a). COTO prescribes only surface slices to be used, requiring multiple cores to be drilled. Hence a conservative DI test result is likely. Interior slices can however be used to assess relatively to the full cured condition.

This module serves as a connecting table between the four earlier modules in *Figure 4-3* Therefore, all identification numbers found in the previous four modules are automatically cross referenced to the current module. The information entered in the previous four modules would typically only have to be entered once, for it to be retained through the reporting of this module. The setup of the physical database in this way enables grouping or filter options for specimens according to the previous four modules. The other relation that is made from the specimen's module is to the test results module which is discussed in the following section.

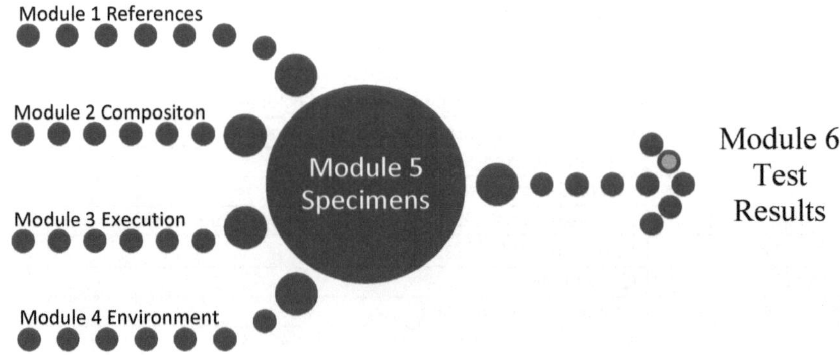

Figure 4-3 Arrangement of the tables within Module 5 (Specimens)

The main table contains the dimensions of the specimen as well as the identification number and its corresponding code for the specimen (*Table 4.26*). Specimens are further distinguished by type, height, width and diameter as indicated in *Table 4-26*. Note all other information contained in *Table 4-26* would be retained from the previous modules through the relevant identification numbers. The preparation of test specimens according to SANS 3001 Part CO3-1: 2015 for concrete DI testing has been used to create the input for this module. A specimen is an object that is tested such as a cube, cylinder or disc and sample is a statistical term for a batch or lot from a mix such as a sample of concrete from which specimens are prepared.

Table 4-26 Main specimen details

Field	Description
Specimen ID	Identification number for the specimen
Specimen code	Identification code for the specimen
CC ID	Identification number for the concrete composition table
Execution ID	Identification number for the execution details table
Environment ID	Identification number for the environment details table
Project ref. ID	Identification number of the project reference table
Specimen	Cube, cylinder, disc etc.
Specimen origin	See list in *Table 4-27*[1]
Date	Date of measurement
Length	Length (mm)
Height	Height (mm)
Width	Width (mm)
Diameter	Diameter (mm)
Name in list	Code for input form (summarised type / dimension)
Condition before	Condition of the specimen upon receipt
Age	Age of concrete calculated from the batch date
Exposure time	Total exposure time for all exposure periods
Curing time	Total curing time for all curing periods
No. of specimens	Total number of specimens prepared
Location of specimen within core / cylinder	Number from 1 to 4 where the former is the outermost surface (away from formwork) and the latter is the innermost surface (against formwork)[1]
Operator	Name of operator responsible for preparation
Add. observations / abnormalities	Comments concerning the specimens 1 to 4 after the testing procedure

Notes: 1   Only of relevance for DI values i.e. mass and compressive strength consists of only one type of specimen. Concrete cover discussed in Section 4.7.6.

Table 4-27 Specimen Origin pre-defined list for input form

Field	Specimen Type
1	Mix design or laboratory (cubes, cylinders or trial panels cast horizontally)
2	Mix design or laboratory (cubes, cylinders or trial panels cast vertically)

3	Construction (test panels cast horizontally)
4	Construction (test panels cast vertically)
5	In-situ structure (cores or discs)

As part of the general requirements according to SANS 3001 Part CO3-1: 2015, cubes are to be cast which are required to be cured in accordance with the specifications not less than 100 mm in dimension. At mix design approval stage, this clause holds true, however, the major alteration occurs before the production stage where "trial" panels are cast, followed by the "test" panels when full production begins on site as part of the quality control scheme in the project specifications. These panels are either cast horizontally or vertically depending on the structural member as indicated in *Table 4-27*. Furthermore, the project specification will dictate the frequency and number of cores per exposed surface area of panel as discussed in Section 3.5.2 (Setting the Sampling Frequency).

It is recommended that the curing method be standardised for all panels since they are cast in field conditions, which differ from site to site. To have some form of consistency, during the "trial" stage, it will be necessary to cast at least two panels for each grade of durability concrete contained within the structure i.e. one set for the superstructure and one set for the substructure. The two panels should contain no curing compound and be water submerged as well as air exposed to represent the extremities of construction quality as accounted for by the relevant DI values. This is such that one can distinguish between pre-qualification or qualification specimens and actual site specimens, the former properly cured and the latter to simulate curing which is not covered in COTO. Additional panels should permit one curing compound per panel which will be cured at the standard application rate, according to the project specifications. Therefore, this method can be used primarily to qualify curing compounds once the mix design has been approved.

SANS 3001 Part CO3-1: 2015 states that specimens (drilled and sliced discs) must contain reference numbers on the interior face with a permanent marker. An important piece of information that needs to be recorded is the order of the specimens from the formwork to the surface. Considering the dimensions of panels, a 150 mm depth core, accounting for the 5 mm recess on either side could possibly provide up to four (30 ± 2) mm specimens that must be recorded to enable such differentiation when analysing the results. The below diagram is suggested for 100 mm depth cores from cubes as indicated in *Figure 4-4*.

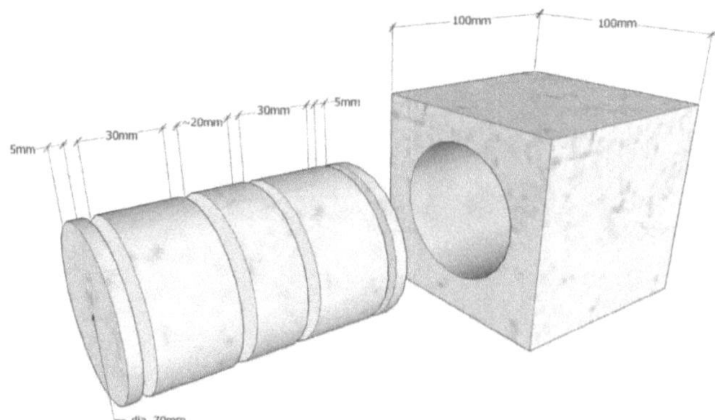

Figure 4-4 Typical core extraction dimensions from cured cubes (SANS 3001-CO3-1:2015)

From actual site concrete elements, coring should only be undertaken if it is not possible to core the existing test panel, when DI values fall into the conditional acceptance, remedial acceptance or rejection categories. In all three cases, coring is essential, since the trend in DI values from test panels to the actual structure depends greatly on the execution and environment for any given concrete composition. The areas with DI values that fall into these categories should be assessed in conjunction with the achieved cover depth to prioritise possible core extraction locations.

Should coring be necessary at more than one location, this must proceed with due caution not to compromise the integrity of the structure and surrounding reinforcement. Therefore, when actual site concrete elements are required to be cored, a method statement should be submitted beforehand to the engineer's satisfaction. The relevant fields to this test method should be specified in Section 4.4 (Module 3: Execution) and Section 4.6 (Module 5: Specimens), in which one should be able to differentiate between the different type of curing (submerged in water, exposed to air and use of curing compounds) and specimen entries i.e. cubes, trial panels, test panels or the actual site concrete elements.

It is explicitly stated that the specimen age may have a significant effect on the test results and for this reason a maximum age should be specified of 56 days. Exceeding this age can further affect other variables such as concrete composition, execution and environment, therefore it is critical that specimens are tested within their allocated time frames for the durability index test procedures (UCT, 2018b). The correlation between the time frame from coring to testing and the DI result using actual data is discussed in Section 5.3.3.1 (Correlation between Specimen Age and DI value) which depends on the parameter under investigation. For mixes containing micro-silica particles or silica fume, a significant amount of microstructural alteration may occur due to the oven drying procedure for high quality concrete and therefore it is pivotal that this type of binder is reported on (UCT, 2018b).

Visser & Han (2003) included separate tables for TEM (resistivity), carbonation, chloride diffusion and migration measurements and stated for statistical analysis that the standard deviation is an important measurement parameter. The material parameter tables have several things in common. The standard deviation will therefore be added to each test method in addition to the variance or COV (%), which has been identified as significant from Section 3.4.3 (Justifying a Maximum Variability or Percentage Defectives). As far as possible, one should be able to cross reference different material tests contained within this module for the different structure types. In saying that, filter options should further allow the possibility breaking down the structure types into smaller portions or grouping them together for entire projects which should be maintained in the reporting of this module discussed in Chapter 5.

Each material test in principle consists of two tables: one containing the actual measurements and one containing details of the measurement equipment and method, such as used apparatus, pre-set clocks and currents and so on (See Appendix D: Entity Relationship Diagram).

The measurement tables have several fields in common, beside the measurement ID number and the specimen ID number referring to the specimens in the table (Specimens) by which all other information is connected, such as the date, the age and the exposure at which the measurement is performed. The age is automatically calculated from the batch date if it is available (stored in the table Concrete Composition), the exposure time has yet to be filled out by hand, since several exposure periods can be supplemented in the table Exposure.

Further, individual measurements are stored in the measurement tables, the amount of fields depending on the measured parameter. For instance, in the case of mass only one weight is recorded per specimen and per date, while for the carbonation depth 12 depth measurements are recorded per specimen and per date. When the amount of measurements per specimens are not known, the measurements themselves are stored in a different table.

The last three fields of each measuring table for statistical analysis consist of the average calculated value, the standard deviation, and the number of specimens in the calculation (note that in the latter case, the specimen code is just a group name). It has been decided to allow for average values in this field because frequently in literature no basic variables are given (such as the weight in the case of the mass and chloride content in the case of diffusion coefficients).

All of the above-mentioned are significant for statistical analysis and probability estimates. Mean values are also calculated for each of the DI test methods since four individual determinations result in one complete result, hence it must be stated that there is a certain degree of increased confidence in the result. Therefore, non-compliance to the specified performance criteria is more pronounced in lots and the exclusion of test results as outliers within samples become less apparent.

4.7.1 Mass

Since mass measurements will always be pivotal in calculation for either carbonation or chloride induced corrosion, where the required DI value is OPI and CCI, it has been decided to standardise this parameter. Note in the case of OPI, the mass measurements are not used, however after specimens undergo OPI testing, the same specimens are commonly tested for WSI as a project specification requirement.

The same holds true for CCI, where specimens can be tested afterwards. Therefore, for at least two different test methods, the same specimen dimensions and resulting mass measurements will be apparent. Mass measurements are also performed for quality control on concrete cubes as well as the TEM, where concrete cubes are stored in different climates (Visser & Han, 2003).

Commonly, following specific mass measurements, the density is calculated from known volume dimensions. The fields for the mass measurement table (*Table 4-28*) and mass detail table (*Table 4-29*) is designed such that no redundant information regarding specimens are stored in the database.

Table 4-28 Mass measurement

Field	Description
Mass ID	Identification code for the mass measurement
Specimen ID	Identification code for the specimen
Mass detail ID	Reference identification number for mass details (as below)
Date	Date of measurement
Mean Weight	Weight (g)
Mean Density	Density (kg/m^3)
Weight std. dev	Standard deviation for weight
Density std. dev	Standard deviation for density
No. of specimens	Number of specimens in the calculation of mean and std. dev

Table 4-29 Mass detail

Field	Description
Mass detail ID	Identification number for mass detail record
Specimen ID	Identification code for the specimen
Comments	Comments concerning the measurement technique

4.7.2 Compressive Strength

As mentioned before, the compressive strength is also pivotal to quality control acceptance procedures similarly to the durability index test methods. The fields for strength measurements is indicated in *Table 4-30* whilst the specific strength detail information can be found in *Table 4-31*. The strength detail is concerned mostly with the set-up of equipment and the treatment of the specimen. It also should be reiterated that compressive strength is prepared in accordance with SANS 3001-CO2-2 and tested in accordance with SANS 3001-CO2-3 where the relevant fields to these test methods are also in *Table 4-30* and *Table 4-31*.

Table 4-30 Strength measurement

Field	Description
Strength ID	Identification code for the strength measurement
Specimen ID	Identification code for the specimen

Strength detail ID	Reference identification number for strength details (as below)
Date	Date of measurement
Mean Force	Maximum load at failure (N)
Mean Strength	Compressive strength (N/mm^2)
Mean Area	Cross-sectional area of specimen (mm^2)
Outlier Check	No. of results that exceed 15 % of average
Force std. dev	Standard deviation for force
Strength std. dev	Standard deviation for strength
No. of specimens	Number of specimens in the calculation of mean and std. dev

Table 4-31 Strength detail

Field	Description
Strength detail ID	Identification number for mass detail record
Specimen ID	Identification code for the specimen
Speed of loading	0.3 MPa/s + 0.1 MPa/s
Preparation method	None, polished or equalised / capped with mortar
Comments	Comments concerning the strength testing technique

4.7.3 Concrete DI testing (Oxygen permeability test – Part CO3-2: 2015 Edition 1)

The fields for OPI measurement are indicated in *Table 4-32*. The OPI detail is found in *Table 4-33* and is concerned mostly with the operator as well as the set-up and calibration of equipment. The specific data to be recorded for the OPI test method can be found in *Table 4-34*. The comments are particularly important for this test method since the test is known to be indicative of macro-structural problems (UCT, 2018b).

Most laboratory ovens are of the forced draft, ventilated type. If, however, the oven being used is of the closed (unventilated) type, then the relative humidity inside the oven must be maintained by the inclusion of trays of saturated calcium chloride solution (UCT, 2018b). The trays should provide a total exposed area of at least 1m^2 per 1m^3 of volume of the oven and should contain sufficient solid calcium chloride to show above the surface of the solution throughout the test (UCT, 2018b). The type of laboratory oven is applicable to all three DI test methods.

Table 4-32 OPI measurement (SANS 3001-CO3-2)

Field	Description
OPI ID	Identification code for the OPI measurement
Specimen Id	Identification code for the specimen
OPI detail Id	Reference identification number for OPI details (as below)
Date oven	Date of OPI specimens in oven
Date desiccator	Date of OPI specimens in desiccator
Date test	Date of OPI measurement test
OPI reading	Final OPI reading (mean of all specimens)
OPI (lot size)	Number of specimens in the calculation of mean

Table 4-33 OPI detail (SANS 3001-CO3-2)

Field	Description
OPI detail ID	Identification number for OPI detail record
Specimen ID	Identification code for the specimen
Operator	Name of operator responsible for OPI test
Oven ID	Equipment number
Oven type	Forced draft ventilated or closed (unventilated)
Oven calibration cert	Calibration certificate (< 5 years old)
Permeability cell ID	Equipment number
Permeability cell calibration cert	Calibration certificate (< 5 years old)
Gauge / transducer type	Electronic or manual
Electronic transducer calibration cert	Calibration certificate (< 5 years old)
Time elapsed	Time frame from coring to testing
Comments	Comments concerning the equipment (any differences from the standard should be stated here)

Table 4-34 OPI data (SANS 3001-CO3-2)

Field	Description
OPI data ID	Identification code for the OPI data
OPI ID	Reference identification code for the OPI measurement
Mean diameter (mm)	Mean diameter of the specimens
Mean thickness (mm)	Mean thickness of the specimens
Cell volume (L)	Cell volume
Z (s^{-1})	Slope of the linear regression line forced through (0,0) point
A (m^2)	Cross sectional area of specimen
T (K)	Absolute temperature in Kelvin
Permeability (m/s)	Coefficient of permeability (k)
r^2	Calculated value of r^2
r^2 validity	Yes if value of $r^2 > 0.99$; No if value of $r^2 < 0.99$
Po (kPa)	Initial pressure at start of test at time to
Time t	List all time intervals in hh:mm:ss
Pt (kPa)	List corresponding pressure measurements at time t
OPI reading (lot)	Final OPI reading (mean of all specimens)
Discarded specimens	Number of specimens discarded for the calculation of mean
Comments	Comments concerning the specimens (including non-compliant r^2 values, visible cracks, honeycombing defects or visible bleed paths)

4.7.4 Concrete DI testing (Chloride conductivity test – Part CO3-3: 2015 Edition 1)

The fields for CCI measurement are indicated in *Table 4-35*. The CCI detail is found in *Table 4-36* and is concerned mostly with the operator as well as the set-up and calibration of equipment. The specific data to be recorded for the CCI test method can be found in *Table 4-37*.

For this test method, there is provision for a retest provided that no longer than 30 minutes has elapsed from completion of the initial test. Therefore, it should be possible to enter a second set of information for the voltage difference (V), electric current (i) and CCI reading (σ). This is one exceptional difference as compared to the OPI test, however it must be reiterated that for both the test methods, the final test reading for the respective test comprises of the mean of 4 individual readings, although this repetition is not shown in the tables.

Table 4-35 CCI measurement (SANS 3001-CO3-3)

Field	Description
CCI ID	Identification code for the CCI measurement
Specimen ID	Identification code for the specimen
CCI detail ID	Reference identification number for CCI details (as below)
Date oven	Date of CCI specimens in oven
Date desiccator	Date of CCI specimens in desiccator
Date test	Date of CCI measurement test
CCI (lot size)	Number of specimens in the calculation of mean

Table 4-36 CCI detail (SANS 3001-CO3-3)

Field	Description
CCI detail ID	Identification number for CCI detail record
Specimen ID	Identification code for the specimen
Operator	Name of operator responsible for CCI test
Oven ID	Equipment number
Oven type	Forced draft ventilated or closed (unventilated)
Oven calibration cert	Calibration certificate (< 5 years old)
Vacuum saturation facility ID	Equipment number
Vacuum saturation facility calibration cert	Calibration certificate (< 5 years old)
Conduction cell arrangement	Simple cell or telescopic tube
Electronic transducer calibration cert	Calibration certificate (< 5 years old)
Time elapsed	Time frame from coring to testing
Comments	Comments concerning the equipment (any differences from the standard should be stated here)

Table 4-37 CCI data (SANS 3001-CO3-3)

Field	Description
CCI data ID	Identification code for the CCI data
CCI ID	Reference identification code for the CCI measurement
Mean diameter (mm)	Mean diameter of the four specimens
Mean thickness (mm)	Mean thickness of the four specimens
Dry mass (g)	Dry Mass (M_d)
Vacuum saturated mass (g)	Vacuum saturated (M_s)
A (m^2)	Cross sectional area of specimen
Voltage difference (V)	Voltage difference (V)
Electric current (i)	Electric current (mA)
CCI reading (lot)	Final CCI reading (mean of all specimens)
Provision for retest	Additional CCI reading (mean of all specimens)
Porosity reading (lot) (%)	Final porosity (n) reading
CCI (lot size)	Number of specimens in the calculation of mean
Discarded specimens	Number of specimens discarded for the calculation of mean
Comments	Comments concerning the specimens (unusual specimen preparation i.e. removal of surface treatment, cracks voids or excessive chipped edges and non-compliant porosity values)

4.7.5 Concrete DI testing (Water sorptivity test – Part CO3-4: Unpublished)

This test method is still to be formalised through the SANS procedure. Nevertheless, it is a frequently reported parameter and due to its relative ease i.e. the same specimens used for OPI and CCI can be used for WSI (only if specimens have not been exposed to moisture from the atmosphere), it is deemed necessary to include in the material tests module. The following tables will display the procedure if one is using new specimens, otherwise, if specimens are used directly from OPI testing, all oven and desiccator details may be omitted, since these will be redundant and invalidate our original database design principles.

The main change is the inclusion of porosity as an important parameter as part of this test method (UCT, 2018b). Even though porosity has always been calculated as part of this test method, this parameter is now just as important in its own right and therefore WSI cannot be viewed in isolation of porosity, since durable concrete should ideally have both low WSI and low porosity values (UCT, 2018b). The correlation between porosity and WSI using actual data is discussed in Section 5.3.2.2 (Correlation between Porosity and WSI) which depends on the quality of concrete and measured mass of specimens. The fields for WSI measurement is indicated in *Table 4-38*. The WSI detail is found in *Table 4-39* and is concerned mostly with the operator as well as the set-up and calibration of equipment. The specific data to be recorded for the WSI test method can be found in *Table 4-40*.

Table 4-38 WSI measurement (SANS 3001-CO3-4 – proposed)

Field	Description
WSI ID	Identification code for the WSI measurement
Specimen ID	Identification code for the specimen
WSI detail ID	Reference identification number for WSI details (as below)
Date oven	Date of WSI specimens in oven
Date desiccator	Date of WSI specimens in desiccator
Date test	Date of WSI measurement test
WSI (lot size)	Number of specimens in the calculation of mean

Table 4-39 WSI detail (SANS 3001-CO3-4 – proposed)

Field	Description
WSI detail ID	Identification number for WSI detail record
Specimen ID	Identification code for the specimen
Operator	Name of operator responsible for WSI test
Oven ID	Equipment number
Oven type	Forced draft ventilated or closed (unventilated)
Oven calibration cert	Calibration certificate (< 5 years old)
Vacuum saturation facility ID	Equipment number
Vacuum saturation facility calibration cert	Calibration certificate (< 5 years old)
Test setup	Paper towels or roller supports
Time elapsed	Time frame from coring to testing
Comments	Comments concerning the equipment (any differences from the standard should be stated here)

Table 4-40 WSI data (SANS 3001-CO3-4 – proposed)

Field	Description
WSI data ID	Identification code for the WSI data
WSI ID	Reference identification code for the WSI measurement
Mean diameter (mm)	Mean diameter of the four specimens
Mean thickness (mm)	Mean thickness of the four specimens
F	Slope of the best fit line from plotting Mwt against square root of hour, in grams
A (m^2)	Cross sectional area of specimen
Msv	Vacuum saturated mass of the specimen
r^2	Calculated value of r^2
r^2 validity	Yes if value of $r^2 > 0.99$; No if value of $r^2 < 0.99$
Mso (kPa)	Initial mass of specimen at time to
Time t	List all time intervals in hh:mm:ss
Mst (kPa)	Mass measurement corresponding to time t
Mwti	Mass gain calculated at interval (Mst - Mso)

WSI reading (lot)	Final WSI reading (mean of all specimens)
Porosity reading (lot) (%)	Final porosity (n) reading
WSI (lot size)	Number of specimens in the calculation of mean
Discarded specimens	Number of specimens discarded for the calculation of mean
Comments	Comments concerning the specimens (including non-compliant r^2 values, non-compliant porosity values, visible cracks, honeycombing defects or visible bleed paths)

4.7.6 Concrete Cover

Concrete cover must be ideally considered as a stochastic variable instead of a constant value and according to fib (2006), five different distribution types are appropriate for the description as well as the variability. Despite the current durability specification's deterministic nature, the definition of DI values and concrete cover according to its mean and standard deviation enable further analysis to be undertaken regarding the stochastic nature of the data.

These would involve assessing the data to decide which distribution types fit the data such as, but not limited to beta, Weibull (min), lognormal and Neville. The latter four distribution types exclude negative values for the concrete cover due to their characteristics whilst the normal distribution does not, however for large concrete covers the normal distribution is very common (fib, 2006). For concrete cover, it is suggested that the mean value should be equal to the nominal value and the standard deviation be equal to 6 mm when additional execution requirements are targeted in project specifications. For restricted distributions the lower limit is equal to 0 mm and the upper limit is equal to 5 times the nominal cover which should be less than the width of the structural element. The fields for cover measurements is indicated in *Table 4-41* which contains the specific data to be recorded whilst the specific cover detail information can be found in *Table 4-42* and is concerned mostly with the operator as well as the set-up and calibration of equipment.

Table 4-41 Cover measurement

Field	Description
Cover ID	Identification code for the cover measurement
Specimen ID	Identification code for the specimen
Cover detail ID	Reference identification number for cover details (as below)
Date	Date of measurement
Total area (m²)	Area of cover survey of lot[1]
Mean cover (mm)	Mean cover of lot[2]
Cover std. dev	Standard deviation of lot for cover
No. of surveys	Number of cover surveys per lot
Mean cover (lot size)	Mean cover of all cover surveys

Notes: 1   *Minimum area of cover survey > 1 m² and minimum of three cover surveys per lot*
       2   *Minimum of 40 individual cover depth readings per square metre (m²) in calculation of mean / std. dev*

Table 4-42 Cover detail

Field	Description
Cover detail ID	Identification number for cover detail record
Specimen ID	Identification code for the specimen
Operator	Name of operator responsible for cover depth test
Device	Cover meter device[1]
Device calibration cert	Calibration certificate (< 5 years old)
Method	Automated[2] or manual[3]
Comments	Comments concerning the measurement technique[4]

Notes:
1   Complying with relevant modern standards (BS 1881 Part 204 and ACI 228)
2   According to the cover metre equipment manufacturer's guidelines
3   Determine position of rebar and manually record readings to establish depth of rebar
4   Should the quick or linear scan method be utilised, additional comments should be stated

4.7.7 Resistivity

For the Two Electrode Method (TEM), resistance is measured between two steel plates, which are compressed to two opposite planes of the concrete cube. At most, two measurements are performed on one cube, since the cast surface is considered to deviate too much from the mould surface to give a reasonable result and is therefore omitted. Based on this method, the table TEM Measurements and TEM detail consists of the fields indicated in *Table 4.43* and *Table 4.44*. There is no table for the calculation method i.e. data since the calculation of resistivity is straightforward and not likely to change.

The specific resistivity res-av is calculated by:

$$Resistance - av\ (Ohm.m)$$

$$= a(scaling\ factor(10^{-3}) \frac{Resistance\ (Ohm)\ x\ Area\ of\ the\ specimen\ (mm^2)}{length\ of\ the\ specimen\ (mm)}$$

Table 4-43 TEM measurement

Field	Description
TEM ID	Identification code for the TEM measurement
Specimen ID	Identification code for the specimen
TEM detail ID	Reference identification number for TEM details (as below)
Date	Date of the measurement
Age	Age of the concrete, calculated from the batch date (if available)
Exposure time	Exposure time of the concrete
Resistance14	Resistance between two opposite planes (Ohm)
Resistance25	Resistance between two other opposite planes (Ohm)
Resistance14	Resistance between two opposite planes (Ohm.m)
Resistancece25	Resistance between two other opposite planes (Ohm.m)
Res-av	Average specific resistivity (Ohm.m)
Res-stdev	Standard deviation of the average specific resistivity
Res-nospec	Number of specimens in the calculation of mean

Table 4-44 TEM detail

Field	Description
TEM detail ID	Identification number for TEM details
Code	Used code
Equipment	Used equipment

4.7.8 Carbonation

Carbonation measurements are performed by splitting specimens at a certain age where the freshly broken surface is sprayed with a chemical substance which colours the carbonated and noncarbonated zone differently. All general details are recorded in the table Carbonation measurement and only the depth is recorded in the table Carbonation depths which consists of the fields as indicated in *Table 4-45* and *Table 4-46*, respectively. These tables are related by the identification code in the former table which appears as a foreign key in the latter table. The Carbonation detail is recorded in *Table 4.47*.

Table 4-45 Carbonation Measurement

Field	Description
Carbo ID	Identification code for the Carbonation measurement
Specimen ID	Identification code for the specimen
Carbo detail ID	Reference identification number for Carbonation details (as below)
Date	Date of the measurement
Age	Age of the concrete, calculated from the batch date (if available)
Exposure time	Exposure time of the concrete
Depth-av	Average carbonation depth (single observation or average calculated from the individual measurements stored in from Carbonation depths (as below)
Depth-stdev	Standard deviation of the average carbonation depth (mm)
Depth-nospec	Number of specimens in the calculation of mean

Table 4-46 Carbonation Depths

Field	Description
Carbo depth ID	Identification code for the Carbonation depth measurement
Carbo ID	Identification code of the specifications of the carbonation measurement from the table – Carbonation Measurement (as above)
Depth	Carbonation depth (mm)

Table 4-47 Carbonation Details

Field	Description
TEM detail ID	Identification code of the carbonation details
Code	Used code
Comment	Comment on the code

4.7.9 Chloride diffusion

Chloride diffusion tests comprises of grinding or crushing layers from a specimen from the exposure surface downwards. Therefore, the chloride content is determined and expressed as either content on cement or on concrete for each depth interval of the concrete material. As for carbonation, it is not known the amount of intervals that will be used, therefore *Table 4-48* and *Table 4-49* are defined for the chloride diffusion test. The former will contain the general information about the test whilst the latter will contain the chloride profiles. The chloride diffusion details are recorded in *Table 4-50*.

Table 4-48 Chloride diffusion measurement

Field	Description
Chloride Diffusion ID	Identification code for the Chloride diffusion measurement
Specimen ID	Identification code for the specimen
Detail ID	Reference identification number for Chloride diffusion details (as below)
Date	Date of the measurement
Age	Age of the concrete, calculated from the batch date (if available)
Exposure time	Exposure time of the concrete
Initial chloride cem	Initial chloride content as weight% of cement
Initial chloride concr	Initial chloride content as weight% of concrete
Start point	Start point of the chloride diffusion calculations
End point	End point of the chloride diffusion calculations
Surf con cem	Surface chloride content in weight% of cement
Diffusion coeff cem	Diffusion coefficient (m^2/s), calculated on cement
Surf conc concr	Surface chloride content in weight% of concrete
Diffusion coeff cem	Diffusion coefficient (m^2/s), calculated on concrete

Table 4-49 Chloride diffusion data

Chloride diffusion data ID	Identification code for the Chloride diffusion depth measurement
Chloride diffusion ID	Identification code of the specifications of the chloride diffusion measurement from the table – Chloride Diffusion Measurement (as above)
Depth Interval	Depth interval on which the chloride content is determined (mm)
Av Depth	Average depth of the interval (mm)
Chloride Cem	Chloride content as weight% of cement
Chloride Concr	Chloride content as weight% of concrete
Calc cem*redundant	Calculated chloride content as weight% of cement
Calc conc	Calculated chloride content as weight% of concrete

Table 4-50 Chloride diff details

Field	Description
Chloride detail ID	Identification code of the chloride diffusion details
Method	Used method
Comment	Any other comments

4.7.10 Chloride migration

The Rapid Chloride Migration (RCM) test is a chloride migration test, where chloride penetration is forced by an electrical current. After the test, the specimen is split and the freshly broken surface is sprayed with a chemical colouring the chloride. Therefore, the penetration depth can be determined. From the chloride penetration front, the chloride migration coefficient is determined.

Like for carbonation and chloride diffusion, the data of this test is split in three different tables: one for all general information concerning the test (RCM Measurements) in *Table 4-51*, one for the penetration depth (RCM Depths) in *Table 4-52* and one for the test specification (RCM Details) in *Table 4-53*. Note that the chloride migration coefficient is calculated according to the Nordic Build Test code.

Table 4-51 RCM measurement

Field	Description
RCM ID	Identification code for the RCM measurement
Specimen ID	Identification code for the specimen
Detail ID	Reference identification number for RCM details (as below)
Start date	Start date of the measurement
Age	Age of the concrete, calculated from the batch date (if available)
Start time	Time at the start of the test
Start temp	Temp at the start of the test
Start potential	Potential
Start currents	Currents
Start resistance	Resistance
End date	Date
End time	Time
End temp	Temp
End current	Current
End resistance	Resistance
co	Chloride concentration in the cathodic solution (≈ 2 in Nordic test)
cd	Chloride concentration at which the colour reaction takes place (=0.07 for the used silver nitrate)
z	Absolute value of the ion valence for chloride (=1 for chloride)
Migration Coeff-av	Average chloride migration coefficient, calculated from RCM Depths
Migration Coeff-stdev	Standard deviation of the corresponding average migration coefficient
RCPT resistivity	RCPT value, calculated from the current during testing resistivity, calculated from the resistance at the start of the test (Ohm.m)

Leakage	Comment whether leakage has taken place and to which extend
Picture	A picture of the penetration front in the specimen

Table 4-52 RCM Depth

Field	Description
RCM depth ID	Identification code for the RCM depth measurement
RCM ID	Identification code of the specifications of the RCM measurement from the table – RCM Measurement (as above)
RCM Depth	Chloride migration depth (mm)
RCM Migration Coeff	Chloride migration coefficient (calculated)

Table 4-53 RCM Details

Field	Description
RCM Detail ID	Identification code of the RCM details
Method	Used method
Comment	Any other comments

4.8 Entity Relationship Diagram (ERD)

An Entity Relationship Diagram (ERD) is a graphical representation of an information system that depicts relationship among concepts within that system. ERD is a data modelling technique that can help define processes and be used as the foundation for a relational database.

The importance of ERDs and their uses:

- ERDs provide a visual starting point for database design

- Used to help determine information system requirements throughout the organisation

- After a relational database is rolled out, the ERD serves as a referral point (debugging or re-engineering needed)

Main components on an ERD:

- ERDs are depicted in one or more of the following models:
  - A conceptual DM: lacks specific detail but provides an overview of the scope of the project and how data sets relate to one another
  - A logical DM: more detailed than a conceptual DM, illustrating specific attributes and relationships among data points
  - A physical DM: While a conceptual DM does not need to be designed before a logical DM, a physical DM is based on a logical DM. A physical DM provides the blueprint for physical manifestation (such as the relational database of the logical DM). One or more physical DM can be developed based on a logical DM

There are three basic components of an entity relationship diagram:

- Entities which are objects or concepts that can have data stored about them

- Attributes which are properties or characteristics of entities. An ERD attribute can be denoted as a primary key, which identifies a unique attribute, or a foreign key, which can be assigned to multiple attributes.

- Relationships between and among entities

There is a lot of moving information in a database and understanding how the many elements of a database interact with each other can be difficult to grasp. This is the reason that engineers rely on a visual way to understand how all these separate elements are related to each other and how they are working together i.e. to build ERDs. The ERD for the DIDb is provided in Appendix B.

Entities are an object such as a person, place or theme to be tracked in the database. Each entity will have attributes which are various properties or traits. In a database, entities will be the rows and attributes will be the columns. Since we have the different entities and attributes, we can now define the relationship that exists between the entities, if they interact with each other at all. This interaction or connection is a relationship in a numerical context defined by a minimum and maximum value called a cardinality as below:

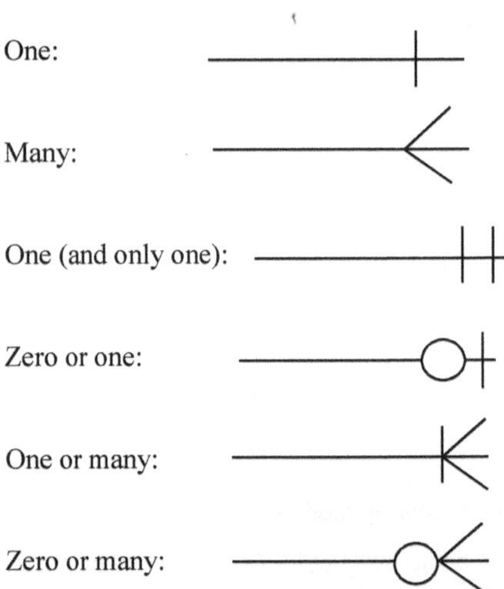

5 Discussion of results

5.1 Introduction

This chapter will focus on identifying relations between different input parameters which adds extra information to the logical data model (DM) elements. Therefore, the relations between the topics are strengthened which ultimately determines the extraction of information or output parameters from the physical database according to specification limits. Five different projects which served as input for a total of 1054 Durability Index (DI) tests were used to conduct parametric studies on the DI values that predominantly affect concrete durability in Reinforced Concrete (RC) structures.

All relevant information defined in the modular structure of the database will be stored in tables defined in Chapter 4. Database design contains four distinct objects that is significant for the database design principles defined in Chapter 3. These are the tools table, query, form and report. A query is a tool or request for data or information from a database table or combination of tables. Queries can be used to answer questions or perform calculations on the data, examples of which will be provided in this chapter. Furthermore, queries can be performed according to specific search or filter criteria that will enable users to group specimens according to the required material test as well as project, concrete composition, execution and environment. The data is generated as results in a report returned by Structured Query Language (SQL).

Visser & Han (2003) state that the major advantage of using queries is that they do not contain any data' itself, but rather codes (ID) which refer to the relevant fields in tables, with formulae which cater for processing and so on. These relevant identification codes (number or name) have also been defined in Chapter 4. An example would be for Module 5 (Specimens), which acts as the central code for the database linking to all other modules and hence all other information.

A query must be performed using pre-defined commands which relate the text input to action. A similar comparison can be drawn with structural analysis software in which users have the option to use the Graphical User Interface (GUI) or edit text field templates. Minimal information must be entered in queries, and often different words will allow users to create various output to suit their needs. Therefore, in the query, only references are stored and data processing will commence once the query has been executed in order to filter, sort or group results from different tables depending on the query requirements. Possible related queries (relationships between different tables within the modules) will be expanded upon in this chapter by referring to output generated for selected projects. Further queries can also be added once the desired and most important relationships are identified (between fields of different tables).

At present, the output should be focused on identifying instances of failure and non-conformance to the specification according to *Table 2-4* as described in this chapter. Furthermore, data should be able to be sub-divided since DI results can vary depending on project and source identified in Section 5.2 (Extraction of Information). The latter refers to the results obtained from trial panels, test panels or in-situ cores.

Therefore, the importance of the parametric study is that it not only compares different projects but also separates DI results according to source in the case of Project 3 and Project 4. This is one of the most important tools that the database should contain which enables the differentiation between trial and test panels in the case of Project 3 and test panels and in-situ cores in the case of Project 4 which ultimately affects the action to be taken in the case of non-compliance with the specification. This chapter concludes with proposing a method to use the numerical summaries of DI test results and the achieved cover depth to calculate the probability that both random variables are out of specification limits in Section 5.3 (Application of 'Deemed-to-satisfy' approach). Five different projects listed in *Table 5-1* have been selected to analyse DI values for conformance with the specification according to the categories provided in *Table 2-4* as well as to provide an indication of the variability. These projects differ in sample sizes as well as composition, execution and environment which further forms the basis for their selection.

Table 5-1 Inventory of data

Project	Name	No. of Results (Determinations)	Period	Short Description
1	N5 Elands River Bridge	18 (72)	23/04/2013 to 06/01/2014	Various bridge elements
2	R35 Amersfoort to Morgenzon	62 (248)	26/03/2013 to 20/06/2014	Major culverts
3	R61 Baziya to Mthatha	33 (132)	21/08/2015 to 25/05/2016	Major culverts
4	N2 Umgeni Interchange	103 (412)	01/07/2011 to 09/07/2014	Bridge substructures, superstructures and culverts
5	N11 Amersfoort to Ermelo	289 (1156)	26/09/2011 to 09/07/2014	Major culverts

1. This project consisted of the realignment of a river bridge to the Harrismith interchange and Kestell in Maluti municipality in the Free State (Moyana, 2015). The section is 2,6 km long with a substantial cutting. In this project a 100 m long in-situ culvert along the Elands River had been installed.

2. This project entailed the Rehabilitation of National Road R35 section 1 from Amersfoort to Morgenzon. SANRAL commenced with road works to upgrade the R35 between Amersfoort and Morgenzon from the month February 2012 and continued for a period of 30 months up to July 2014. The project involved the improvement, rehabilitation and strengthening of the existing road, adding climbing and passing lanes and the upgrading of the intersections (Moyana, 2015).

3. This project entailed the Upgrade of National Route 61 section 7 from Baziya to Mthatha. The client was SANRAL and the contract duration was 36 months. The scope involved the widening to a 13,4 m surfaced width, additional auxiliary lanes and 14 major culverts (Moyana, 2015).

4. The Umgeni Interchange was a split diamond interchange which was unable to accommodate the then existing traffic demand. It was replaced with a four-level system interchange of two grade separated directional ramps that cross the N2 by means of viaducts (Moyana, 2015). Two directional overpass structures were placed parallel to the N2 crossing the M19 of 70 m length. Directional loop ramps and five on and off ramps were also added as part of this project.

5. This project entailed the rehabilitation of the National Route N11 from Amersfoot to Ermelo in Mpumalanga. The project included provision of climbing lanes, upgrade of storm water structures and the upgrading of intersections of a 49 km single carriageway (Moyana, 2015).

5.2   Extraction of Information (Output)

5.2.1 Concrete Durability Specification Limits

A system of classes, safety margins and targets currently exist with respect to the "potential" durability of RC structures in relation to DIs as a function of the environmental exposure classes as defined in EN206-1 and target service life (SANRAL, 2009).

The assessment of DI values for test panels has proven to characterise important variables encountered in the field that govern durability performance which include the type and extent of curing as well as compaction and bleeding effects. In comparison to standard laboratory moist cured conditions, the sensitivity and variability of the results needs to be assessed according to the adopted construction regime to identify its impact on the early age development properties of concrete. High sources of variability and non-conformance with the specification should be highlighted for the main purpose of improving the quality of construction and eliminate poor practice within projects. The secondary purpose can be to link the non-conformance to possible causes and assess the influence on medium or long-term concrete durability performance. This can be done by collating the DI data that contains information regarding different material, manufacturing and testing conditions, defining these parameters more accurately in terms of their mean and standard deviation in degradation models which ultimately form the basis from which a service life prediction can be made.

In these mathematical models, DIs and monitoring parameters are involved as input and output data, respectively, however the nature of the vast amount of data needs to be understood, before it can be utilised more effectively and profitably in industry. The performance-based approach relates environmental classes to quantitative exposure categories which combined with the required OPI or CCI results can compute estimates of the carbonation depth and chloride concentration for a given cover. This process is complex depending on other interrelated parameters such as the concrete composition and execution accounted for by the DI values and the environment which are possible sources of variability. This approach using nominal DI values does not take into account the variability and oversimplifies the matrix of parameters affecting concrete durability (Moyana, 2015). However, the more comprehensive and complex application of the DI approach, involves assessing the main influencing parameters that affect durability of a specific structure within a specified environment which is known as the rigorous approach.

The parameters involved in the rigorous approach include concrete composition (Module 2), execution (Module 3), which correlates to the curing method (OPI value) and cover to reinforcement (specified against achieved), environment (Module 4), which correlates to the exposure conditions and the notional design life which is selected as a 100-years.

In COTO (2018a), for the purposes of durability (D-class) concrete, it is stated that structures require an extended service life of 100 years in typical environments that require a minimum of 80 % of the service life to be free from the risk of corrosion. This condition is provided that the nominal DI values specified are attained according to the design assumptions, however this performance-based (stochastic) specification which refers to the deemed-to-satisfy (deterministic) approach to DI testing can create confusion on how to analyse results for conformance.

5.2.2 Verification of Durability Specification

The deemed-to-satisfy approach has also changed substantially since its inception in 2009 depending on refinements in the DI values. A comparison between the initial and current specification can be found in Appendix D as well as a verification in the case of the latter for the different environmental classes. The comparison revealed that the OPI specification has become more lenient while the CCI specification has become stricter. The results obtained from evaluating the deterioration model in terms of OPI and CCI for both upper and lower limits showed the following characteristics. For OPI, the lower limit is critical, as expected and is associated with reduced safety factors for the cover depth. The cover depth is exceeded for 10 different conditions for the lower limit as opposed to 5 in the case of the upper limit. These conditions are summarised per environmental class in *Table 5-2* and *Table 5-3*. For CCI, the upper limit is critical, as expected and is also associated with reduced safety factors for the cover depth. The cover depth is exceeded for 2 conditions which is provided in *Table 5-4*.

Table 5-2 Carbonation Depths for specified nominal OPI value (100-year design service life)

Environmental Class	Environment Category	OPI Limit (log scale)	Concrete Composition	Specified Cover (mm)	Carbonation Depth (mm)
XC1a	20 - Coastal	9.15	FA / SF	40	47.7
		9.00	FA / SF	50	53.0
XC3	10 – Dry inland	9.65	FA / SF	40	41.2
		9.35	FA / SF	50	55.7
		9.05	FA / SF	60	70.3

Table 5-3 Carbonation Depths for minimum permissible OPI value (100-year design service life)

Environmental Class	Environment Category	OPI Limit (log scale)	Concrete Composition	Specified Cover (mm)	Carbonation Depth (mm)
XC1a	20 - Coastal	8.90	PC / BS / CS	40	41.0
		8.90	FA / SF	40	56.5
		8.75	FA / SF	40	61.8
		8.75	FA / SF	50	61.8
XC3	10 – Dry inland	9.40	FA / SF	40	53.3
		9.10	FA / SF	50	67.8
		8.80	FA / SF	60	82.4
XC4	10 – Dry inland (Wetting – drying)	9.60	FA / SF	40	43.6
		9.30	FA / SF	50	58.1
		9.05	FA / SF	60	70.3

Table 5-4 Threshold Chloride Content (0.4 %) Depth for upper CCI limit (100-year design service life)

Environmental Class	Environment Category	CCI Limit	Concrete Composition	Specified Cover (mm)	Chloride Depth (mm)
XS1	30 - Severe	0.60	SF	50	55.0
		0.85	SF	60	64.0

Typically, the OPI or CCI requirements are specified in terms of nominal values which have a lower and upper limit, respectively. The judgement in accordance with the specification for OPI or CCI depends on the acceptance limit for the parameters of 0.25 (Log scale) and 0.40 (Milli Siemens/cm), respectively which indicates the maximum permissible deviations for which no conditional acceptance is applied. However, should values exceed this limit, then rejection limits of 0.40 for both OPI and CCI are proposed, respectively to further classify the data under the remedial acceptance or rejection categories.

This classification implies that there is variability contained within the data that should be verified during or after construction. Therefore, inspections must be undertaken to evaluate conformity within the design data for actions and fib (2006) states that the planned activities on inspection should focus on the evaluation of the design data applied in deterioration models.

Therefore, the following section will focus on a parametric study for the DI values. According to fib (2006), should the inspection or monitoring reveal that the original service life design assumptions are not met, then five different categories of action should be taken.

For new construction, repairing or strengthening the structure to bring performance back to the agreed design assumptions and protecting the structure to reduce the action is often undesirable, however the latter may be warranted in some instances when nominal DI values deviate greatly from the specification. This also depends on the actual cover achieved as opposed to the specified nominal cover. The last option according to fib (2006) is that the structure should become obsolete, which is applicable to existing structures, however before this option is considered, other steps must be considered. Such steps involve widening the scope of the performance survey to improve the quality and representativeness of the data or performing a recalculation of the original Service Life Design (SLD) to assess the residual service life of the structure. Furthermore, fib (2006) states that the new calculation shall be supplemented with the data for action, materials and products derived from the field-exposed structure including that the redesign conforms to the requirements for the basis of design. Repair, strengthening or protection of the structure must be based on either a partial or full recalculation of the original service life design assumptions to assess the residual service life of the structure according to fib (2006). Therefore, the recording of DI values in a database not only improves the quality and representativeness of the data but also supplements designers with the data for action derived from the field exposed structure.

5.3 Application of 'Deemed-to-satisfy' approach (to EN206)

The parametric study involved transforming the data according to the Z-score normalisation process designated as Method 1 in COTO (2018b) and discussed in Section 2.4.3 (Defining Outliers). Phi (Ø) which was defined as approximately 10 % according to the margins and confidence levels stated in Section 2.4.3 (Defining Outliers) was also checked and compared to the actual percentage defectives. The mean DI value which relates to the specification categories and CoV which relates to the repeatability data according to Section 3.5.4 (Lab Equipment Used) was also compared.

The specification as discussed in Section 2.4.1 (COTO Concrete Durability Specification) will be used in terms of its sample mean ($\bar{X}n$) and lower or upper acceptance limit (La or L'a) represented by the blue and red vertical lines in the below three figures, respectively and applied to DI results obtained from a construction site testing various bridge elements. Therefore, this example will act as a preliminary analysis for the remaining four projects assessed in Chapter 5. The output presented in this analysis is in the form of parameter numerical summaries (tables), data representations (according to acceptance categories) and plots of the parameter standard normal distributions which are suggested to be adopted for the extraction of information or output parameters for the physical database.

5.3.1 N5 Elands River Bridge

Data was obtained from 18 results (72 determinations) that were taken on cores from test panels representative of bridge elements. The data included testing for OPI and WSI as well as 11 tests for CCI. The dates of casting ranged from 23/04/2013 to 06/01/2014. The dates of testing samples were also reported and the periods from casting to testing were all within acceptable margins, ranging from 28 to 29 days. The numerical summary for the data is indicated in *Table 5-5*.

Table 5-5 Parameter numerical summary (Project 1)

Parameter	Minimum	Maximum	Range	Mean	Standard Deviation	Within CoV (%)
OPI (Log scale)	8.74	10.12	1.38	9.43	0.43	4.52
OPI CoV (%)	0.57	5.19	4.62	1.98	1.23	-
WSI (Mm/hour$^{0.5}$)	4.21	11.85	7.64	7.36	2.21	29.99
WSI CoV (%)	3.97	63.01	59.03	17.15	12.82	-
CCI (Milli Siemens/cm)	0.51	2.28	1.77	1.29	0.68	52.79
CCI CoV (%)	1.57	4.24	2.66	3.24	0.88	-

The mean CoV of 1.98 % for OPI falls within the acceptable range as per the repeatability standards of between 1.50 % to 3.00 % for site data, however the within CoV of 4.52 % greatly exceeds this range and is the highest for all projects. The standard deviation for the CoV is 1.23 % and 16.67 % of OPI values exceed the maximum allowable percentage of 3.00 %.

The mean CoV of 3.24 % for CCI falls below the acceptable range as per the repeatability standards of 10.00 % to 15.00 % for site data, however the within CoV of 52.79 % greatly exceeds this range and is the highest for all projects. The standard deviation for the CoV is 0.88 % and no CCI values exceed the maximum allowable percentage of 15.00 %.

It should be noted that the within CoV for WSI of 29.99 % is also the highest for all projects, apart from one dataset in a particular project. The highest CoV for WSI was recorded for cores extracted from the actual structure, which proves that the additional field variability due to the material, manufacturing and testing conditions varies the most in these circumstances.

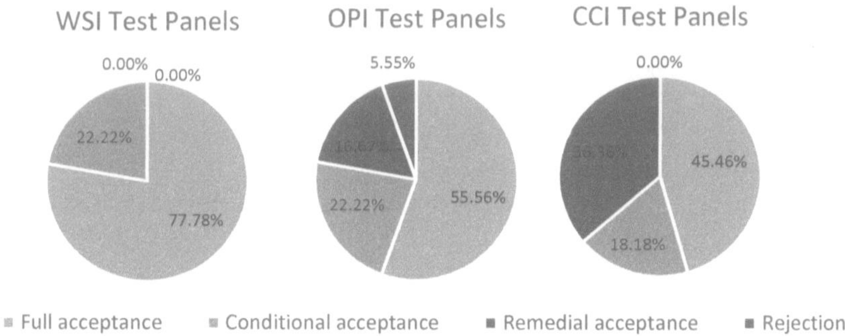

Figure 5-1 Data representation for Project 1

From *Figure 5.1* and *Figure 5.2*, by evaluating the lower acceptance limit (La), the amount of defectives for OPI is equivalent to 44.44 %. From *Figure 5.1* and *Figure 5.3*, by evaluating the upper acceptance limit (L'a), the amount of defectives for WSI is equivalent to 22.22 %. From *Figure 5.1* and *Figure 5.4*, by evaluating the upper acceptance limit (L'a), the amount of defectives for CCI is equivalent to 54.54 %.

Therefore, all DI parameters exceed the 10 % limit proposed by Alexander, Ballim, & Stanish (2008) and conditional acceptance should be further investigated. Furthermore, the lower or upper rejection limit (Lr or L'r) must be evaluated to determine whether remedial acceptance and/or rejection is also applicable. In terms of CCI, it should be noted that the specification is also greatly dependent on the type of binder used, and hence this information should be captured from different projects.

However, in current specifications, such as COTO (2018b), there is no criteria for remedial acceptance and rejection. For instance, considering remedial acceptance, whether the fixed payment adjustment factors as for conditional acceptance are applied which is discussed in Section 2.4.5 (Justifying a Maximum Variability or Percentage Defectives), how the defective concrete should be remedied and, probably the most pertinent of all questions, how to ascertain if the desired performance has been met to reinstate full payment after the remedial work has been carried out. For rejection, it is also not clear what further measures will be taken, what payment will be made nor whether the defective concrete shall be removed and replaced.

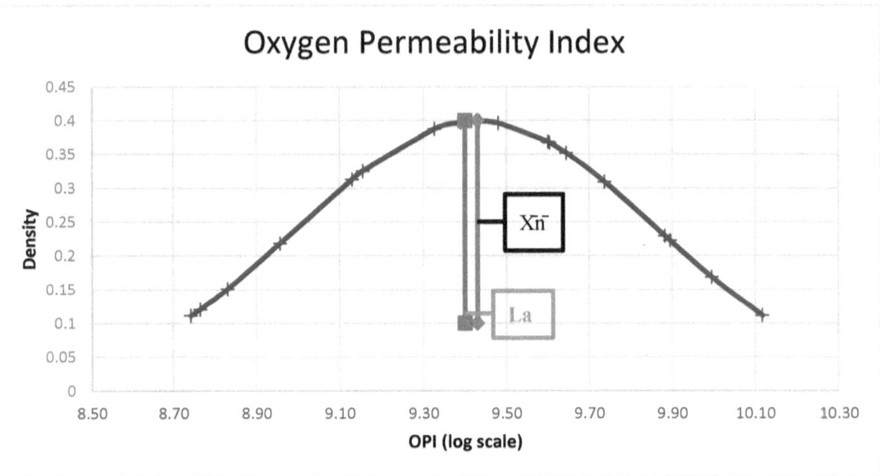

Figure 5-2 OPI Standard Normal Distribution (Project 1)

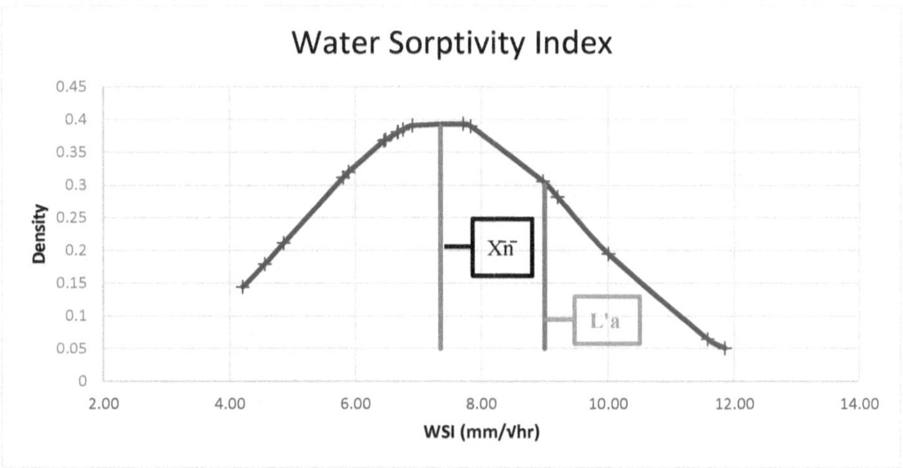

Figure 5-3 WSI Standard Normal Distribution (Project 1)

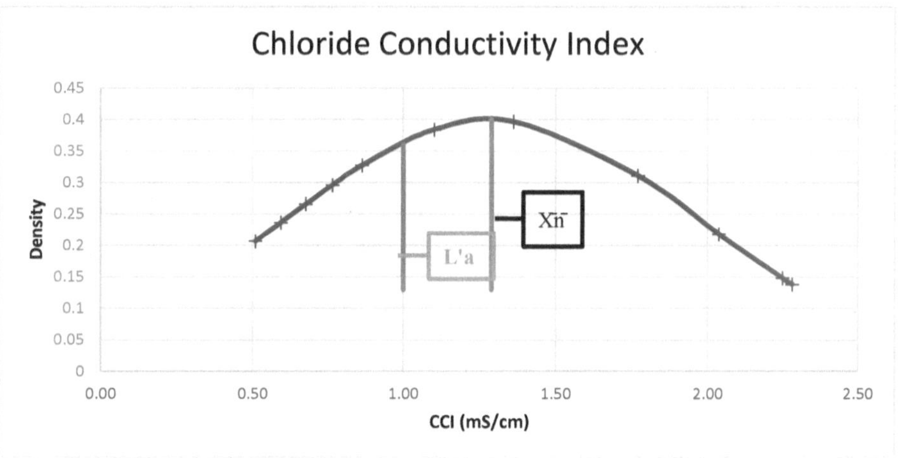

Figure 5-4 CCI Standard Normal Distribution (Project 1)

5.3.2 R35 Amersfoort to Morgenzon

Data was obtained from 62 results (248 determinations) that were taken on cores from test panels. The data included testing on OPI and WSI and also included other additional and mandatory parameters in Module 6 (Test Results) such as Permeability and Porosity values further discussed in Section 5.3.2.1 (Correlation between Permeability and Carbonation) and Section 5.3.2.2 (Correlation between Porosity and WSI), respectively. Only the date of sample delivery was reported which ranged from 26/03/2013 to 20/06/2014 hence no information regarding the sample age could be inferred. The numerical summary for the data is indicated in *Table 5-6*.

Table 5-6 Parameter numerical summary (Project 2)

Parameter	Minimum	Maximum	Range	Mean	Standard Deviation	Within CoV (%)
OPI AVG	8.90	10.08	1.18	9.36	0.22	2.33
OPI CoV (%)	0.20	5.29	5.09	1.62	1.05	-
k AVG	8.41E-11	2.11E-09	2.02E-09	5.26E-10	2.90E-10	55.09
k CoV (%)	4.56	128.33	123.78	34.67	22.60	-
WSI AVG	4.52	14.23	9.71	9.34	1.97	21.13
WSI CoV (%)	2.30	58.19	55.89	12.13	8.88	-
n AVG	5.35	15.63	10.27	10.91	2.44	22.32
n CoV (%)	1.85	31.65	29.80	9.60	5.82	-

The mean and within CoV for OPI of 1.62 % and 2.33 %, respectively falls within the acceptable range as per the repeatability standards of between 1.50 % to 3.00 % for site data. The standard deviation for the mean between CoV is 1.05 % and 8.06 % of OPI values exceed the maximum allowable percentage of 3.00 %.

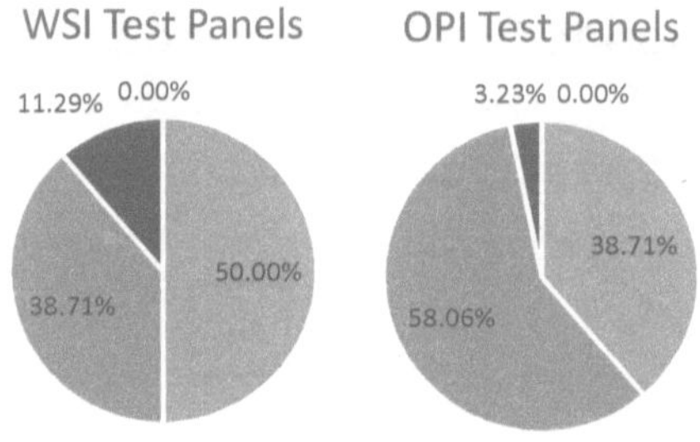

Figure 5-5 Test Panel data representation (Project 2)

According to *Figure 5.5*, the number of defectives that fall within the conditional acceptance and remedial action category account 50 % for WSI and 61.29 % for OPI. This percentage equals the number of specimens that fall within the full acceptance category for WSI and exceeds that for OPI.

5.3.2.1 Correlation between Permeability and Carbonation

The Oxygen Permeability Index (OPI) correlates well with the rate of carbonation, which is affected by material, manufacturing and testing conditions. More specifically, the depth of carbonation depends on the concrete pore geometry, size, interconnectedness and the chemical nature of the binder.

In fly ash or slag blended cements, effective curing is of particular importance. The amount of carbonatable material in the form of calcium hydroxide (Ca[OH]$_2$) is considerably less in unblended cements resulting in higher carbonation rates in any given environment. Furthermore, the hydration reactions of most blended cements are much slower than plain Portland cement implying that longer duration curing regimes are necessary to achieve an equivalently dense pore structure. Literature indicates OPI is sensitive to slight variations in composition (w/b ratio) as well as curing and compaction hence it can serve as a good indicator of concrete quality.

The reporting of additional parameters allows for further analysis to be taken and is recommended to be implemented for each of the tests contained within Module 6 (Test Results). Examples consist of defining parameters in terms of their Standard Normal Distribution to assess the variability as outlined in Section 2.4 (Quality Control Scheme for Concrete Durability). The Standard Normal Distribution of permeability which is used to calculate the OPI value is indicated in *Figure 5.6*.

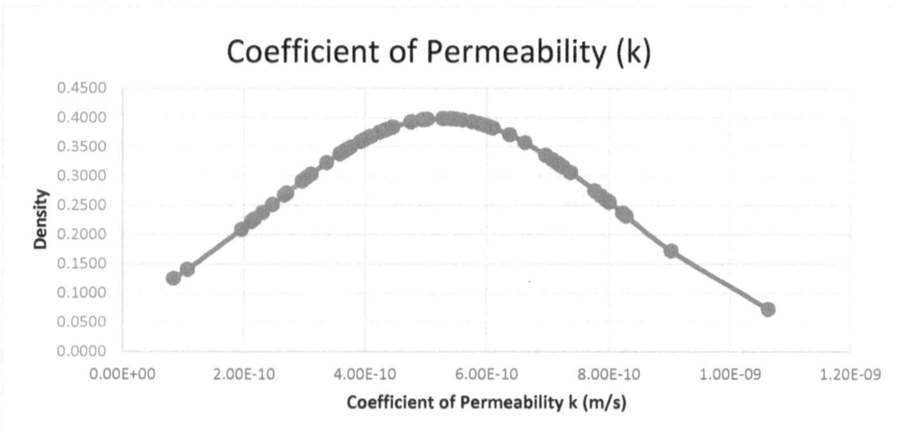

Figure 5-6 Permeability Standard Normal Distribution (Project 2)

The CoV for Permeability ranges from 4.56 % to 128.33 %. The mean and standard deviation of the CoV is 34.67 % and 22.60 %, respectively. The within CoV for the data is 55.09 % which is in the caution range. This value also exceeds the repeatability values for site data which ideally ranges from 40 – 50 %. The minimum and maximum permeability values differ by a factor of 25, in which the latter value implies that concrete is 25 times more permeable than the former. These permeability values also correspond to OPI values of 8.90 (log) and 10.08 (log), which indicates that an increase in magnitude of 1 or more on the OPI scale has a significant effect on the permeability of concrete.

It should be noted that in all of the projects analysed, the range of the OPI value exceeded 1, with the exception of trial panels in Project 3. This can be attributed to the fact that specimens were assessed under trial conditions in the laboratory where standard moist curing practices were used.

The range of OPI values can hence be used as a good indicator for execution relating to different curing regimes. Low OPI values on average can be compared to shorter or ineffective curing practices, whilst high OPI values will be associated with longer curing periods. The permeability coefficient of concrete can also be related to the effective diffusion coefficient which can be used further to calculate the depth of carbonation using the first and second equation as below (Gopinath, Alexander, & Beushausen, 2014). Also referred to as the carbonation coefficient, this parameter depends on the relative humidity of the environment and the OPI value. Salvoldi (2010) also proposed a humidity factor H_s to account for the influence on relative humidity on carbonation as indicated by the last equation.

$$C = \sqrt{\frac{2D_c}{a}} \times \sqrt{t}$$

$$D_c = mk^n H_s$$

$$H_s = 23.32 \, (1 - [RH/100])^2 ([RH/100])^{2.6}$$

RH_{real} data for Project 2 based on trends in the relative humidity for a period of 10 years correspond to an average (mean) daily value of 57.51 % which corresponds to an inland environment and Hs of 0.999. This environment also corresponds to the highest carbonation coefficient in comparison to coastal and partly wet environments since the most favourable exposure condition for carbonation is between 50 % and 70 % relative humidity (Gopinath, Alexander, & Beushausen, 2014). Using the empirical constants of 126 and 0.96 for parameters m and n, respectively which were based on data from the OPI test and natural carbonation test for Portland Cement (PC) and Fly Ash (FA) samples, proposed by Gopinath, Alexander, & Beushausen (2014), the relationship between the permeability and diffusion coefficient can be estimated for Project 2 indicated in *Figure 5.7*. Therefore, with further information regarding concrete composition, required input parameters from physical database can be used to validate such factors or empirical constants and even extended to calculate the depth of carbonation and apply service life predictions to different projects.

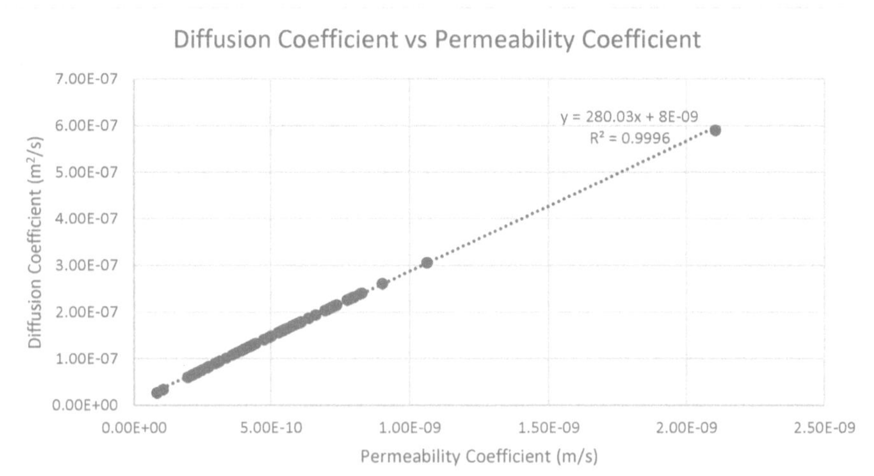

Figure 5-7 Correlation between Diffusion Coefficient and Permeability Coefficient (Project 2)

5.3.2.2 Correlation between Porosity and WSI

The Water Sorptivity Index (WSI) assesses the rate of absorption of water uni-directionally into a concrete medium. This primarily occurs due to capillary action of the concrete pores and depends on the pore geometry as well as the degree of saturation. WSI is very sensitive to near surface transport properties and presents with it increased variability in comparison to OPI. However, when specimens are wet cured, the variability of the test decreases quite substantially which proves the test method is strongly dependent on construction factors, such as the degree of curing and methods of finishing for concrete, and hence has the potential to be used effectively as a site control parameter. It is stated in COTO (2018b) that for conventional (normal-density) concrete, the porosity of the specimen shall be greater than 6 % in order for the test to be considered as valid. In general, the results presented exceed this minimum limit according to *Figure 5-8*. Good porosity values can range from 8 % to 12 % and therefore when this percentage is exceeded, poor quality concrete can be expected. The CoV for Porosity ranges from 1.85 % to 31.65 %. The mean and standard deviation of the CoV is 9.60 % and 5.82 %, respectively. The within CoV for the data is 22.32 % which is comparable to that of WSI.

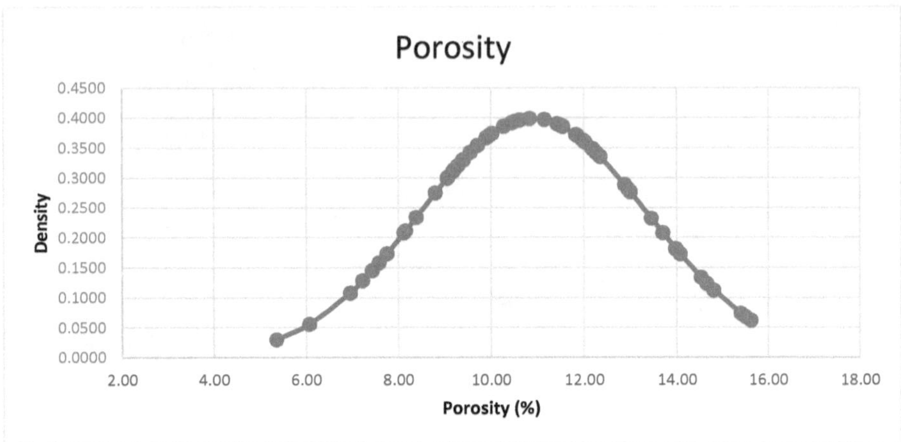

Figure 5-8 Porosity Standard Normal Distribution (Project 2)

The reporting of the additional information, which is mandatory furthermore, allow trends to be analysed between various parameters. *Figure 5-9* indicates that with increasing WSI, porosity also increases. This trend occurs for WSI and porosity values between 4.52 mm/√hr to 14.23 mm/√hr and 5.35 % to 15.63 %, respectively. With further information regarding the mass of specimens, it would be possible to refine these trends even further for different ranges of WSI and porosity and increase the subsequent R^2 value. These results also show the importance of assessing both WSI and porosity in relation to each other during analysis, since good concrete should ideally display both a low WSI and porosity.

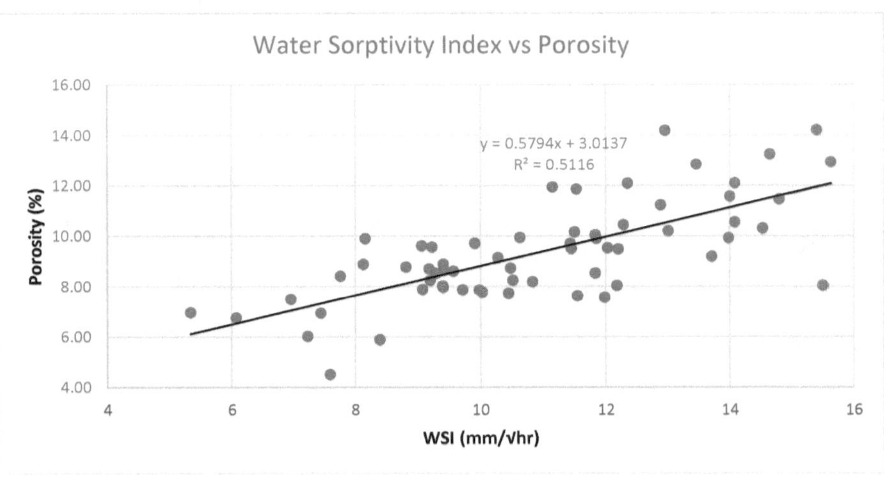

Figure 5-9 Correlation between WSI and Porosity (Project 2)

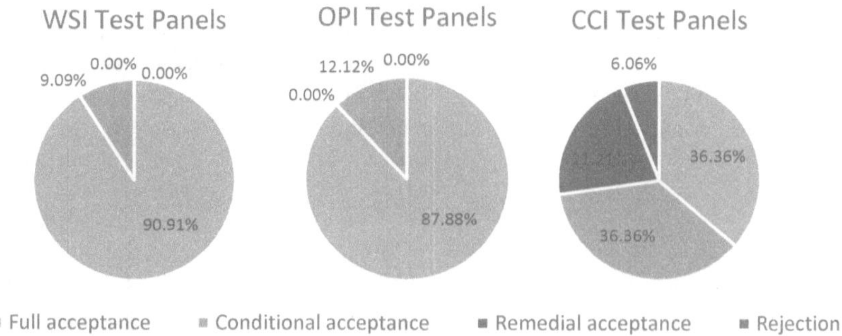

Figure 5-10 Test Panel data representation (Project 3)

5.3.3.1 Correlation between Specimen Age and DI value

The DI tests are to be conducted on early age samples of concrete, which are cored from specimens approximately 26 to 32 days from casting. When this time frame is exceeded, the reliability of the DI values can be questioned such that poor concrete may appear to perform better due to the development and maturity of the concrete microstructure with time.

Even though the trend is not clearly visible with OPI values since they are measured on the logarithm scale, the increase in sample age clearly produces better DI values for both WSI and CCI, in which lower values are more desirable, irrespective of the material, manufacturing and testing conditions (*Figure 5-11*). Therefore, in such instances, the DI values need to be assessed and margins need to be appropriately adjusted to cater for this ageing effect of specimens. Typically, values that appear to meet the DI specification need to be reduced for OPI and increased for WSI and CCI, whilst values that do not the DI specification after prolonged periods from casting to testing would typically worsen if tested under early age conditions.

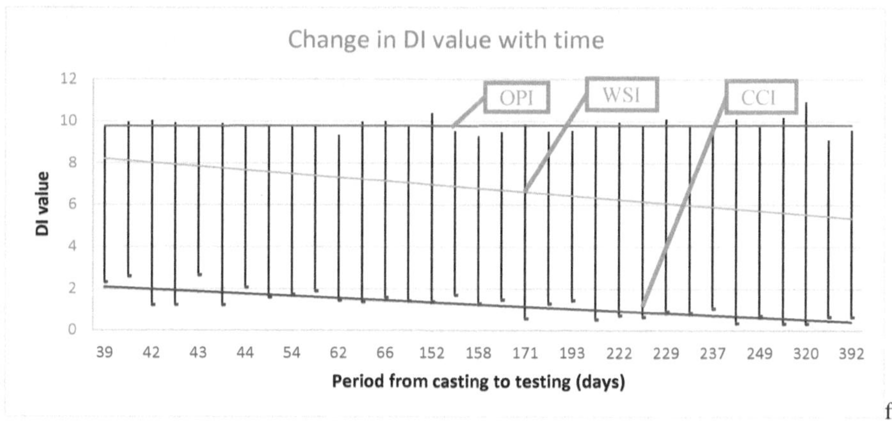

Figure 5-11 Trends in DI values with different periods from casting to testing (Project 3)

It should however be noted there is no apparent trend in the Coefficient of Variation (COV) for either DI test, which proves that a prolonged period from casting to testing does not cancel out the variability in the specimen (*Figure* 5-12). Therefore, the material, manufacturing and testing conditions still play an important role and is information that is captured by DI parameters even despite the prolonged period from casting to testing.

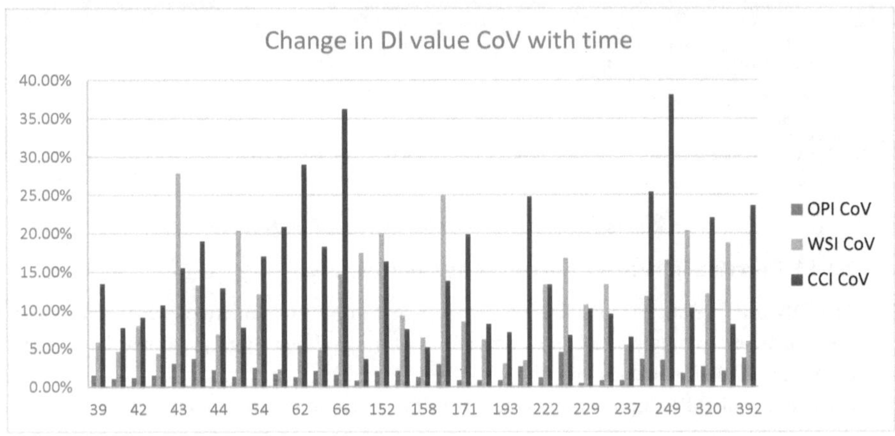

Figure 5-12 Trends in CoV values with different periods from casting to testing (Project 3)

5.3.3.2 Comparison between Project 2 and Project 3

A comparison between Project 2 and Project 3 indicates the following. From *Figure 5-13*, even though the within CoV for Project 2 is smaller than that of Project 3, due to the reduction of the mean OPI value, more defectives are present in Project 2. Therefore, even with the increased variability present in Project 3, OPI values still fall above target values. From *Figure 5-14*, the within CoV for Project 2 is greater than that of Project 3. Furthermore, an apparent increase of the mean WSI value results in more defectives present in Project 2. For Project 3, the within CoV and mean WSI values are reduced which is indicative of better quality concrete.

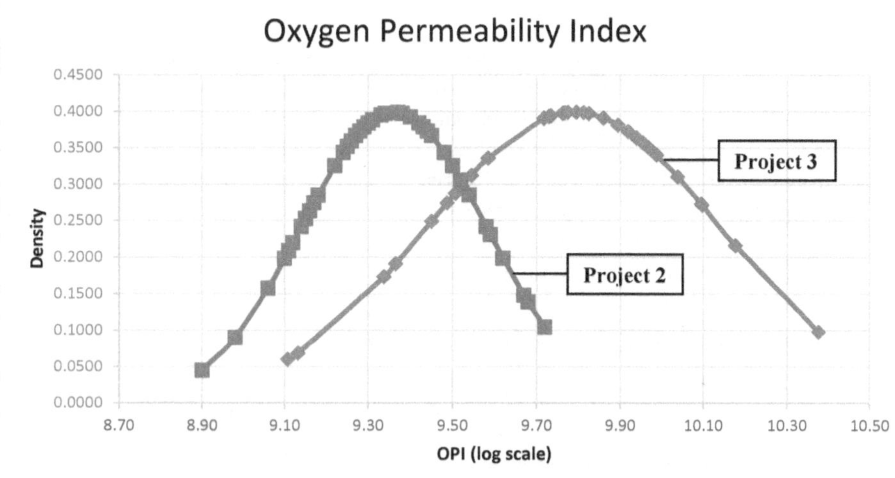

Figure 5-13 OPI Standard Normal Distribution (Project 2 and Project 3)

For Project 2, the mean OPI value of 9.36 (log) and CoV of 2.33 % results in a high number of defectives of 61.29 %, whereas for Project 3, the mean OPI value of 9.79 (log) and CoV of 3.60 % results in only 12.12 % defectives. For Project 2, the mean WSI value of 9.34 mm/√hr and CoV of 21.13 % also results in a high number of defectives of 50.00 %, whereas for Project 3, the mean WSI value of 6.78 mm/√hr and CoV of 22.85 % results in only 12.12 % defectives. Therefore, an increased CoV does not always result in increased defectives but should be assessed in line with the mean value. This proves the importance of assessing both mean values and the within CoV in projects to get an accurate reflection of the DI results.

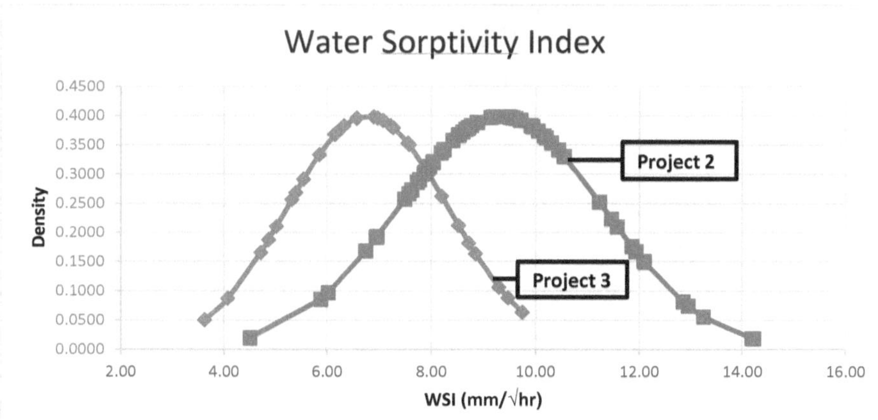

Figure 5-14 WSI Standard Normal Distribution (Project 2 and Project 3)

5.3.4 N2 Umgeni Interchange

Data was obtained from 78 results (312 determinations) that were taken on cores from test panels representative of bridge substructures, superstructures and culverts. The data included testing on OPI and WSI, however a substantial amount of trial results were submitted in relation to other projects. Trial testing occurs under laboratory conditions and the remainder of the results were taken on cores from test panels on site. For trial panels, both dates of casting and testing samples were reported which were all within acceptable margins, ranging from 34 to 39 days. For test panels, only the date of sample delivery was reported which ranged from 01/07/2011 to 09/07/2014 hence no information regarding the sample age could be inferred. The numerical summary for the data is indicated in *Table 5-8*.

Table 5-8 Parameter numerical summary (Project 4)

Parameter	Minimum	Maximum	Range	Mean	Standard Deviation	Within CoV (%)
OPI$_{TR}$ AVG	10.03	10.98	0.95	10.57	0.23	2.13
OPI$_{TE}$ AVG	9.15	10.59	1.44	9.94	0.29	2.87
OPI$_{TR}$ CoV (%)	0.69	5.00	4.31	1.78	1.15	-
OPI$_{TE}$ CoV (%)	0.24	6.20	5.95	2.15	1.32	-
WSI$_{TR}$ AVG	3.98	7.75	3.77	5.97	1.10	18.48
WSI$_{TE}$ AVG	2.89	10.58	7.69	5.54	1.57	28.25
WSI$_{TR}$ CoV (%)	4.22	29.88	25.66	12.74	6.63	-
WSI$_{TE}$ CoV (%)	2.48	37.19	34.71	12.58	7.10	-

The mean value for OPI reduces when considering trial panels in relation to test panels. However, the mean value for WSI also reduces in the same circumstances. The mean CoV and within CoV for OPI increases when considering trial panels to test panels. However, the mean CoV reduces and within CoV increases for WSI in the same circumstances. Therefore, it is possible to achieve better quality concrete, in terms of WSI, even with an increased within CoV. According to *Figure 5-15* and *Figure 5-16*, all parameters are within the full acceptance category for WSI with only 3.77 % defectives and 5.66 % defectives for OPI.

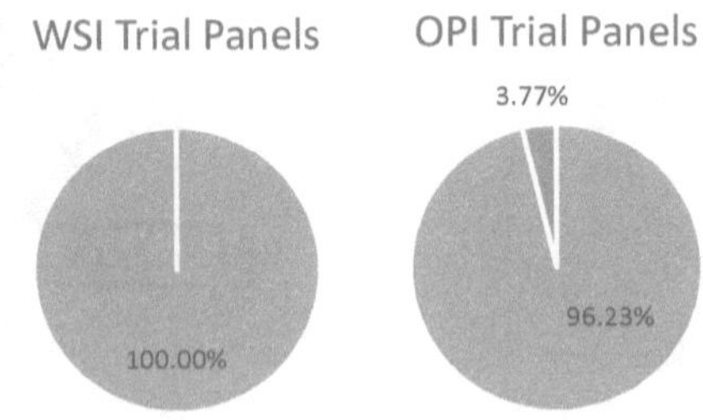

Figure 5-15 Trial Panel data representation (Project 4)

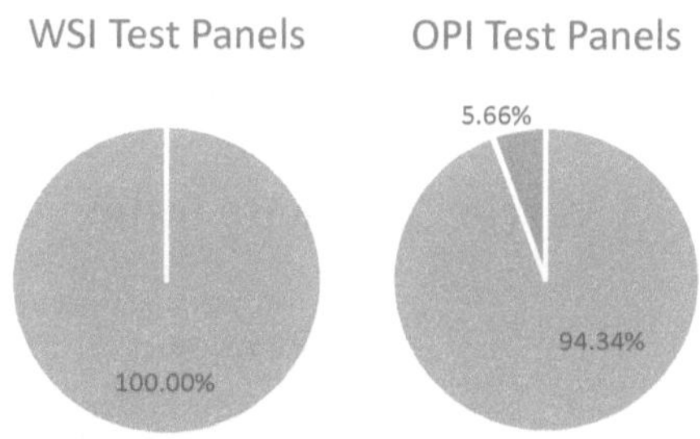

Figure 5-16 Test Panel data representation (Project 4)

5.3.4.1 Comparison between Trial Panels and Test Panels

A comparison between trial panels and test panels indicates the following. From *Figure 5-17*, the within CoV increases for test panels as compared to trial panels, as expected. Since trial panels are cast under laboratory conditions and test panels are cast in the field, additional variability is attributed to the material, manufacturing and testing conditions. The mean OPI value also shifts from 10.57 (log) to 9.94 (log) when considering trial panels in relation to test panels.

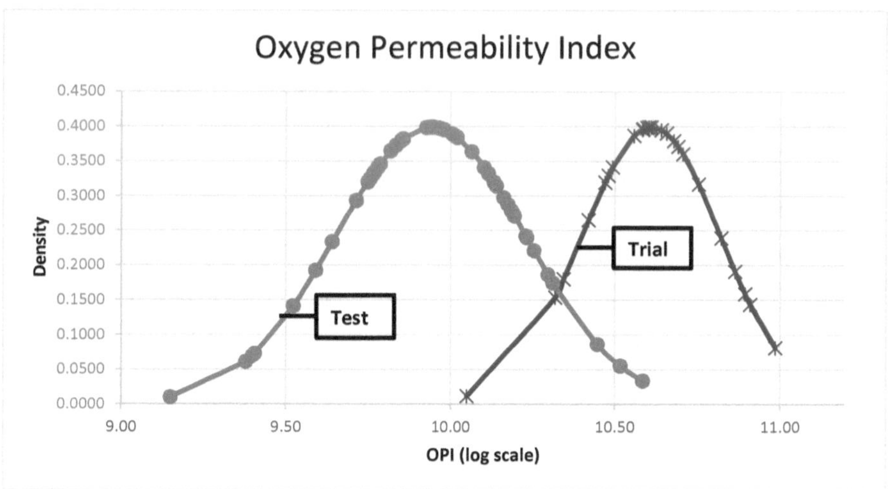

Figure 5-17 OPI Standard Normal Distribution (Project 4)

From *Figure 5-18*, the mean WSI value decreases from 5.97 mm/√hr to 5.54 mm/√hr when considering trial panels in relation to test panels. This proves that superior results can be obtained under test conditions in the field compared to trial conditions in the laboratory with good construction practices. Although this decrease in WSI is small, considering the sensitive nature of this DI parameter in comparison to OPI, the more notable difference is the increase in within CoV of 9.77 % which can be attributed to the field conditions.

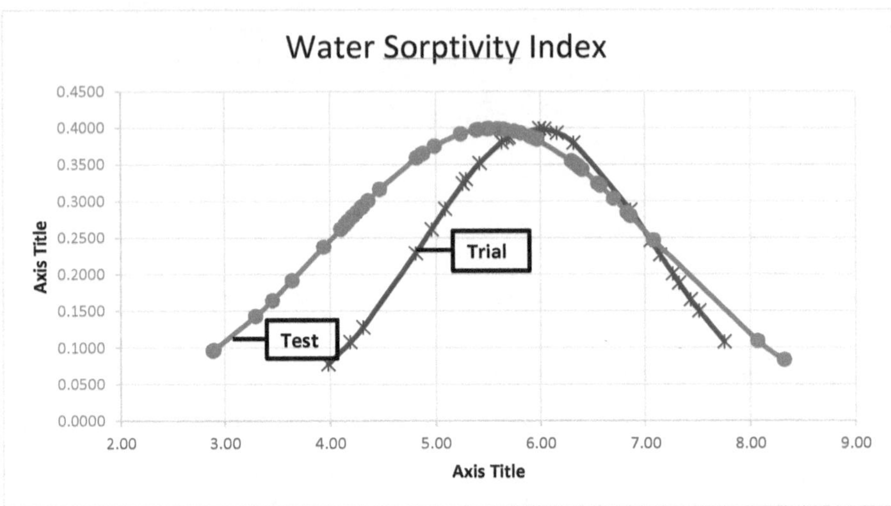

Figure 5-18 WSI Standard Normal Distribution (Project 4)

5.3.5 N11 Amersfoort to Ermelo

Data was obtained from 261 results (1044 determinations) that were taken on cores from test panels representative of major culverts. The data included testing on OPI and WSI, which included other additional and parameters such as the corresponding compressive strength as well as results obtained from cores extracted from the actual structure. Only the date of sample delivery was reported which ranged from 26/09/2011 to 09/07/2014 hence no information regarding the sample age could be inferred. The numerical summary for the data is indicated in *Table 5-9*.

Table 5-9 Parameter numerical summary (Project 5)

Parameter	Minimum	Maximum	Range	Mean	Standard Deviation	Within CoV (%)
OPI$_P$ AVG	8.67	10.57	1.90	9.67	0.29	3.05
OPI$_C$ AVG	8.54	9.73	1.19	9.26	0.31	3.39
OPI$_P$ CoV (%)	0.10	7.41	7.31	1.78	1.13	-
OPI$_C$ CoV (%)	0.18	9.83	9.65	2.22	1.83	-
WSI$_P$ AVG	2.82	14.12	11.30	7.16	1.67	23.36
WSI$_C$ AVG	4.56	17.70	13.14	8.46	3.16	37.36
WSI$_P$ CoV (%)	0.44	56.65	56.21	13.16	8.10	-
WSI$_C$ CoV (%)	0.29	42.20	41.91	16.11	8.96	-

The mean value for OPI reduces when considering panels in relation to cores and the mean value for WSI increases in the same circumstances. The same trend can be identified in both the CoV and within CoV, as expected with the additional variability attributed to the field conditions.

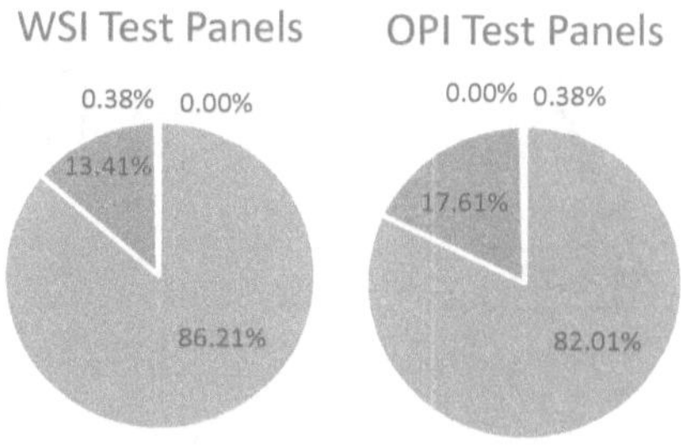

Figure 5-19 Test Panel data representation (Project 5)

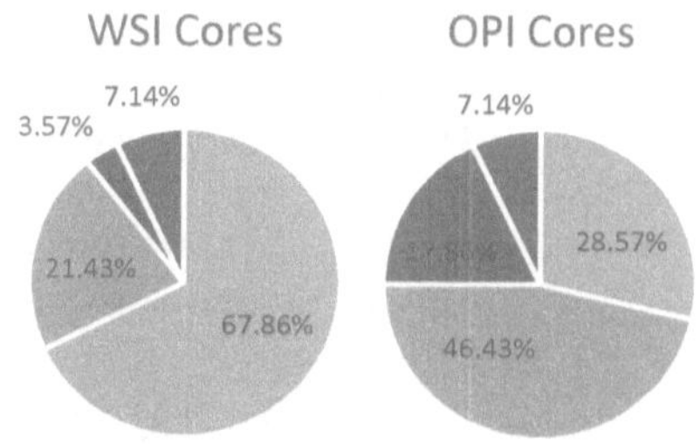

Figure 5-20 Core data representation (Project 5)

According to *Figure 5-19* and *Figure 5-20*, the full acceptance categories decrease when considering test panels in relation to cores for both DI parameters. The conditional acceptance categories increased from 17.61 % to 46.43 % for OPI and 13.41 % to 21.43 % for WSI. Whilst minimal values were considered to fall within the remedial action and rejection categories when considering test panels, cores produced defectives falling into these categories of 25 % for OPI and 10.71 % for WSI.

5.3.5.1 Comparison between Test Panels and Cores

A comparison between test panels and cores indicates the following. From *Figure 5-21*, the within CoV can be closely related for both test panels and cores, which supports the fact that there is constant material, manufacturing and testing conditions in the as-built structure and test panels. However, the mean OPI value does shift when considering test panels in relation to cores from 9.67 (log) to 9.26 (log).

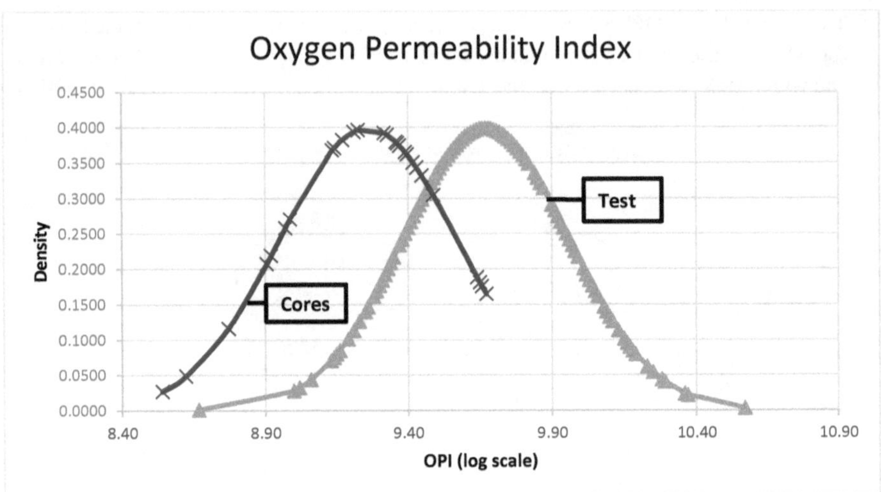

Figure 5-21 OPI Standard Normal Distribution (Project 5)

From *Figure 5-22*, there are increases in both the mean value and within CoV for WSI when considering test panels relation to cores, which supports the fact that this DI parameter is more sensitive to material, manufacturing and testing conditions in the as-built structure as compared to test panels. The increase in mean WSI value from 7.16 mm/√hr to 8.46 mm/√hr is also accompanied by an increase in within CoV of 14%. For cores extracted from the actual structure, the WSI within CoV value is the highest as compared to all other projects.

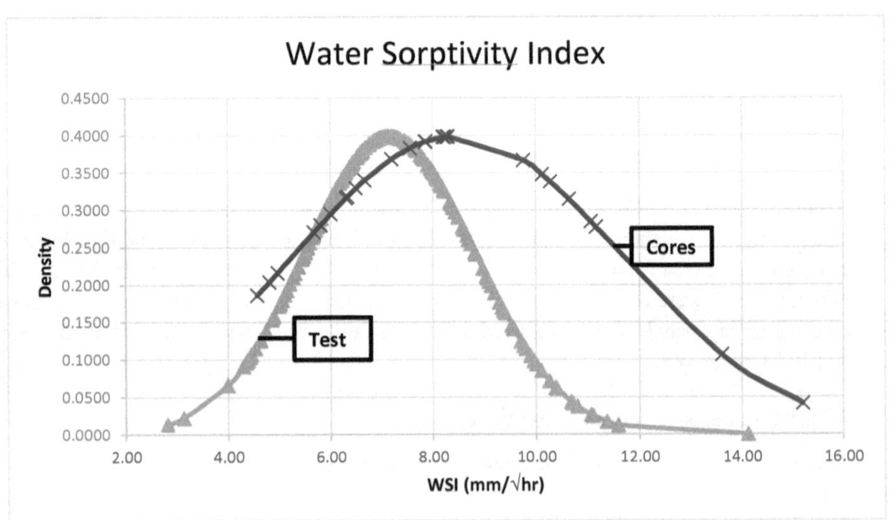

Figure 5-22 WSI Standard Normal Distribution (Project 5)

Since the test panels had corresponding results for cores extracted from the actual structure, further correlational analysis was possible for OPI, WSI as well as their respective CoV. This analysis also keeps the material, manufacturing and testing conditions constant between the as-built structure and test panels. From *Figure 5-23*, only 10.71 % of the OPI values tested from cores were higher than that from test panels, whilst the majority of 89.29 % of OPI values were all lower. This trend is also repeated for WSI, in which only 28.57 % of the WSI values tested from cores were lower than that from test panels, whilst the majority of 71.43 % of WSI values were all higher (*Figure 5-24*). This proves that cores found within the actual structure can be of a poorer quality than those from test panels, in terms of OPI, WSI and their respective CoV.

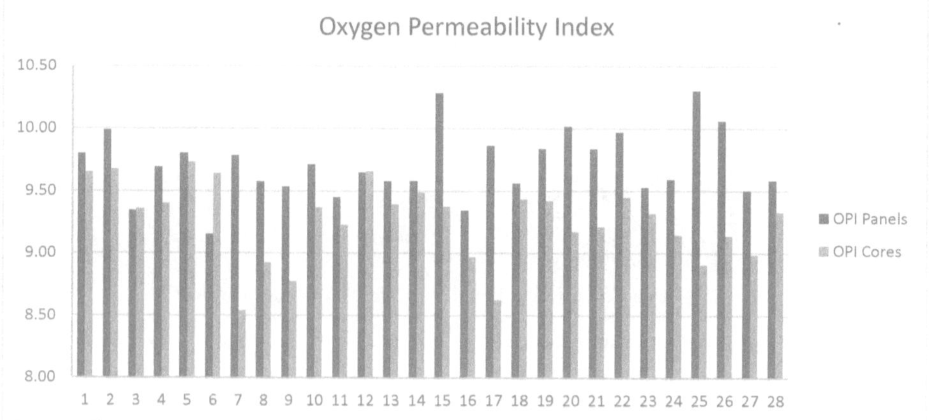

Figure 5-23 OPI Test Panel Vs Core values (Project 5)

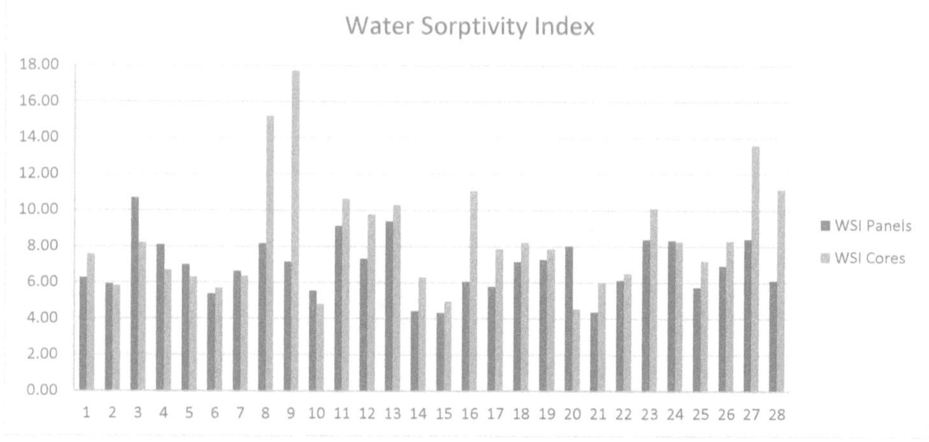

Figure 5-24 WSI Test Panel Vs Core values (Project 5)

Regarding the CoV, there seems to be no apparent trend between test panels and cores. From

Figure 5-25, 40.74 % of the OPI values tested from cores had a lower CoV than that from test panels, whilst the remaining 59.26 % all had a higher CoV. The same holds true for WSI, in which 46.43 % of the WSI values tested from cores had a lower CoV than that from test panels, whilst the remaining 53.57 % all had a higher CoV (*Figure 5-26*).

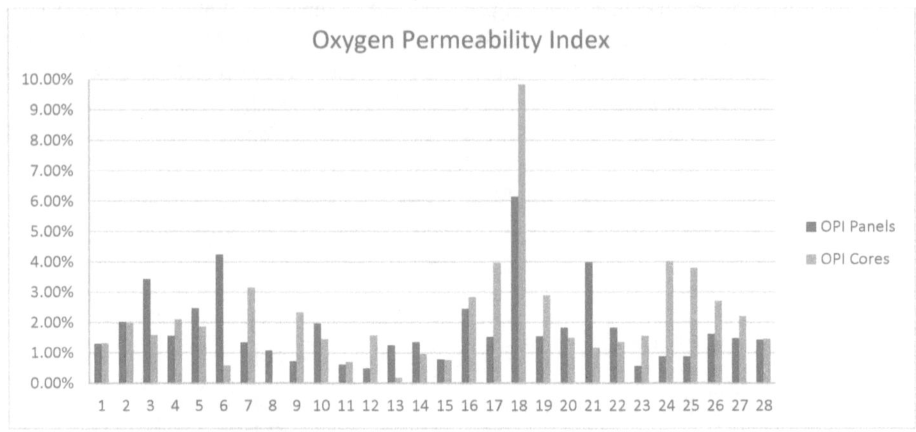

Figure 5-25 OPI Test Panel Vs Core CoV values (Project 5)

However, upon further examination, cores (3, 6 and 12) which displayed higher OPI values in relation to test panels, contained a lower CoV on 2 out of 3 occasions, which indicates that higher quality is associated with reduced variability. However, cores (2, 3, 4, 5, 7, 10, 20 and 24) which displayed lower WSI values in relation to test panels, contained a higher CoV on 6 out of 8 occasions, which indicates that higher quality can also be associated with increased variability. This proves the importance of assessing the mean value, CoV and relevant DI parameter to ascertain the quality from the as-built structure or represented by test panels.

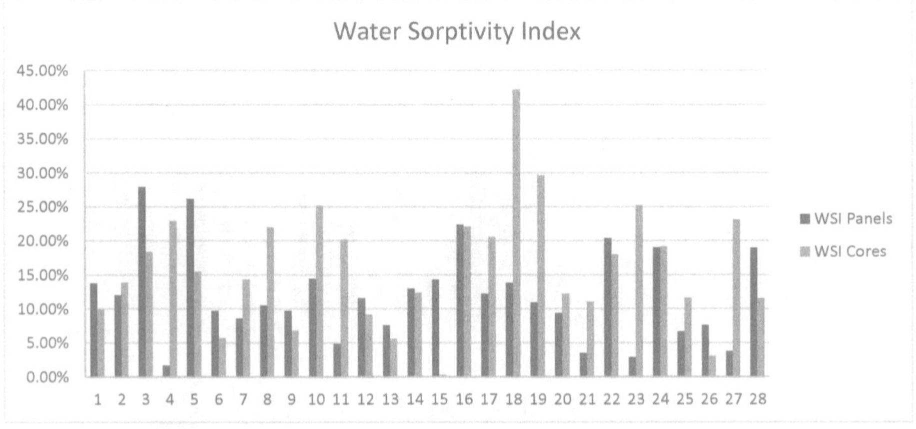

Figure 5-26 WSI Test Panel Vs Core CoV values (Project 5)

5.4 Application of Rigorous Approach (to EN206)

Assessing DI values and cover depth in terms of their Standard Normal Distribution allows one to assess the variability for the advantages outlined in Section 2.4 (Quality Control Scheme for Concrete Durability) but most importantly to sufficiently conclude the occurrence or probability that either value is in accordance with the specification.

These correlations regarding the data can be referred to as conventional which allow the characterizing of a percentage defectives or the probability of obtaining DI values below the limiting value. OPI will be further used as an example to undertake conventional and knowledge-based probability calculations for the different projects. The results obtained in *Table 5-10* show that larger sample sizes increase the reliability of the calculation, but this is not always the case.

Table 5-10 Probability of OPI < 9.40 (Project 1 – 5)

	Project 1	Project 2	Project 3	Project 4		Project 5	
	TEST	TEST	TEST	TRIAL	TEST	TEST	CORES
Results	18	62	33	25	78	261	28
Determinations (x4)	72	248	132	100	312	1044	112
Mean OPI	9.43	9.36	9.79	10.57	9.94	9.67	9.26
Standard Deviation	0.43	0.22	0.35	0.23	0.29	0.29	0.31
CoV (%)	4.52	2.33	3.60	2.13	2.87	3.05	3.39
Defectives (%)	44.44	61.29	12.12	3.77	5.66	17.61	71.43
Z-score	-0.07	0.19	-1.11	-5.76	-1.86	-0.93	-0.45
P (Z)	0.5279	0.5753	0.8643	1.000	0.9686	0.8238	0.6736
P (OPI < 9.40)	0.4721	0.4247	0.1357	0.000	0.0314	0.1762	0.3264

For instance, the probability calculation approximately equals the percentage defectives in the case of test panels for Project 5 which contains the greatest number of specimens. This also proves that the OPI data is well modelled by a standard normal distribution when considering large amounts of data and therefore can be assumed to follow this distribution in other instances for the purposes of data analysis.

However, Project 4 illustrated that the margin between trial and test panel OPI results is 0.63 whereas in Project 5 the margin between test panel and core OPI results is 0.41 which exceeds both the acceptance and rejection limits as outlined in Section 2.4.3 (Defining Outliers). The latter case (Project 5) is more critical whereas the former case (Project 4) contains DI values within the specification, however both cases prove the importance of isolating DI results according to source (trial panels, test panels or cores), which greatly influences the analysis for conformity with the specification.

The conventional probability calculations generally correspond with the percentage defectives in all occasions except for two. This occurs in Project 2 (test panels) and Project 5 (cores) which also represent the two occurrences of the highest percentage of defectives of 61.29 % and 71.43 %, respectively. In all cases the percentage defectives for OPI should be an ideal indicator to further examine other parameters such as the cover depth since it is linked to a degradation model which can be used to evaluate the carbonation depth.

5.4.1 Durability Index (DI) Values & Cover Depth Readings

The risk of corrosion depends on not only the measured OPI but also on the achieved cover. Therefore, assuming that OPI values and cover depth are two random variables that are normally distributed allows correlations to be more expert or knowledge-based where questions can be answered in terms of probabilities for more than one criterion or conditions.

Both parameters (OPI values and cover depth) defined in terms of their mean and standard deviation and assuming independence between parameters allows one to plot the Joint Probability Density Function using the normally distributed values. Therefore, if we designate the required pairs of OPI value and cover depth as (x, y), one can compute the probability that any value of (x, y) is in a specific region by determining the volume over that region.

An example is illustrated using OPI data from Project 2 and Project 3 which has been defined in terms of its mean and standard deviation. It has been assumed that the mean cover depth is equal to 50 mm and the standard deviation is equal to 6 mm for each case, whilst the limiting value for OPI is 9.40 (log) in line with the specification categories *Table 2-4*. It should be noted that cover can show greater standard deviation than the value assumed. Where post-casting control of cover is actively measured that the mean cover may be +2 mm to +4 mm greater than specified. Where cover is retrospectively measured, significant reductions in mean cover and increased variability of cover typically result.

The volume of the region OPI < 9.40 (log) and cover depth < 50 mm will be equivalent to the probability of obtaining a set of values satisfying both these conditions. From *Figure 5-30* and *Figure 5-31*, one can clearly see that the volume of the region concerned is significantly greater for Project 2, as expected.

To determine the volume under a three-dimensional surface chart based on data following a non-linear relationship, integration calculations are required to be performed for the equation describing the surface (5th or 6th order curve or other) which is a complex procedure. Therefore, simplifications are necessary to arrive at reasonable probability estimates for joint low OPI values and cover depth. Applying the cumulative function to the normally distributed values allows this type of Joint Probability Density Function to converge to 1 as indicated in firstly in *Figure 5-27*, secondly in *Figure 5-28* and thirdly in *Figure 5-29*. Therefore, the error and probability of combined low OPI values and cover depth can be determined by simple calculation using parameters defined in terms of their mean and standard deviation and by using the below .XLS spreadsheet in Table 5-11 to plot *Figure 5-30, Figure 5-31, Figure 5.32* and *Figure 5-33*.

Table 5-11 Percentage error (accepted value and experimental value)

OPI	0.175			% ERROR
COVER	0.5			
JOINT	0.0875	experimental value		0.0005
		% error	= av-ev	=0.005682
			=|av-ev|	% ERROR
OPI	0.1759		av	
COVER	0.5			
JOINT	0.088	accepted value		

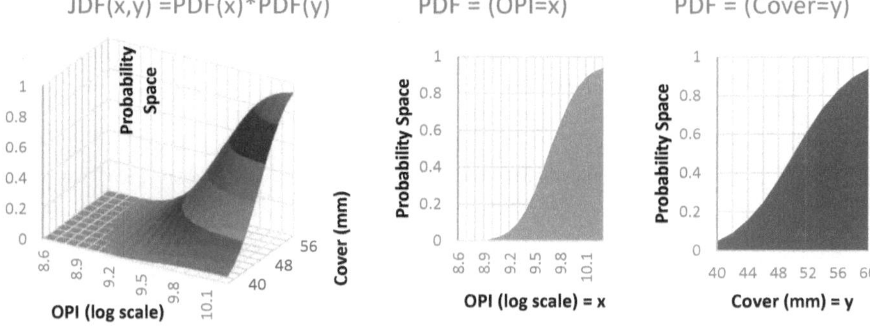

Figure 5-27 JDF (Project 5)   Figure 5-28 PDF (OPI)   Figure 5-29 PDF (Cover)

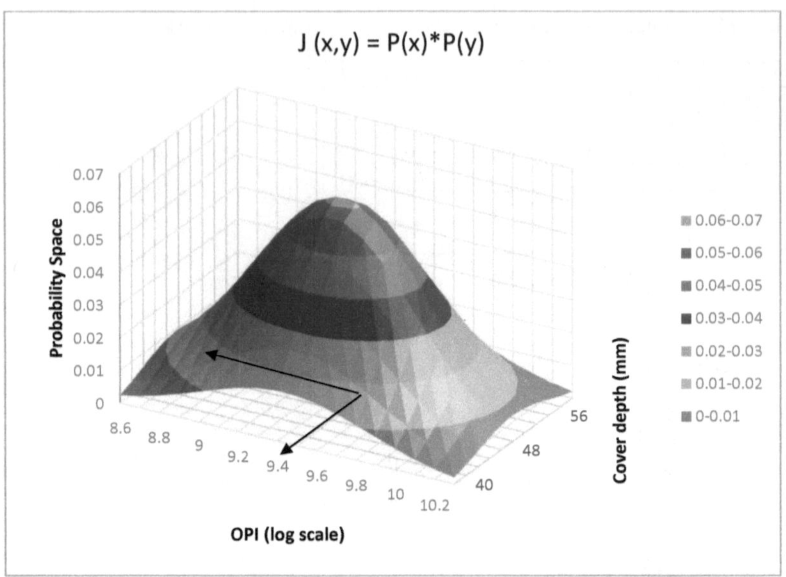

Figure 5-30 Joint Probability Density Function (Project 2)

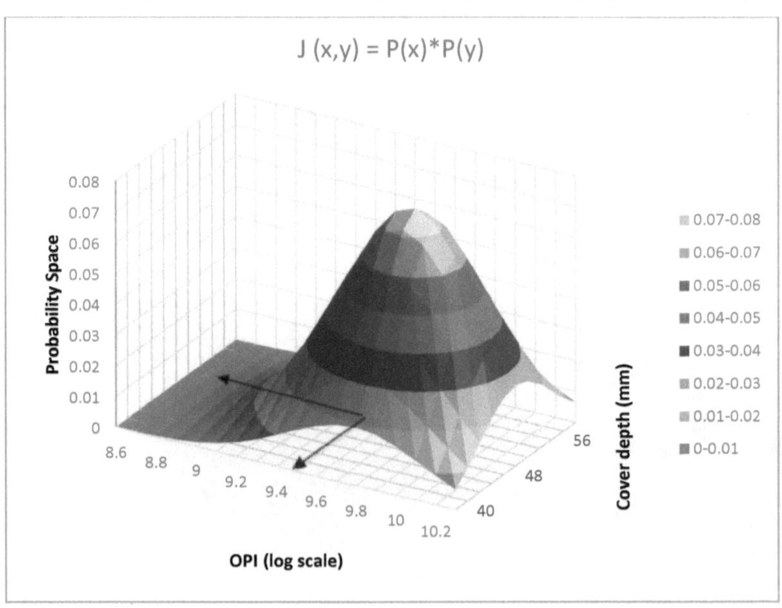

Figure 5-31 Joint Probability Density Function (Project 3)

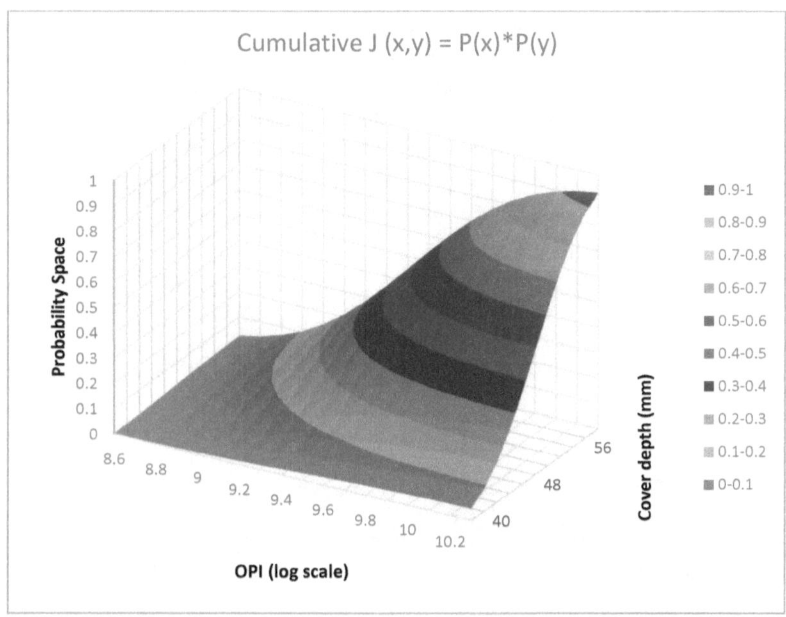

Figure 5-32 Cumulative Joint Probability Density Function (Project 2)

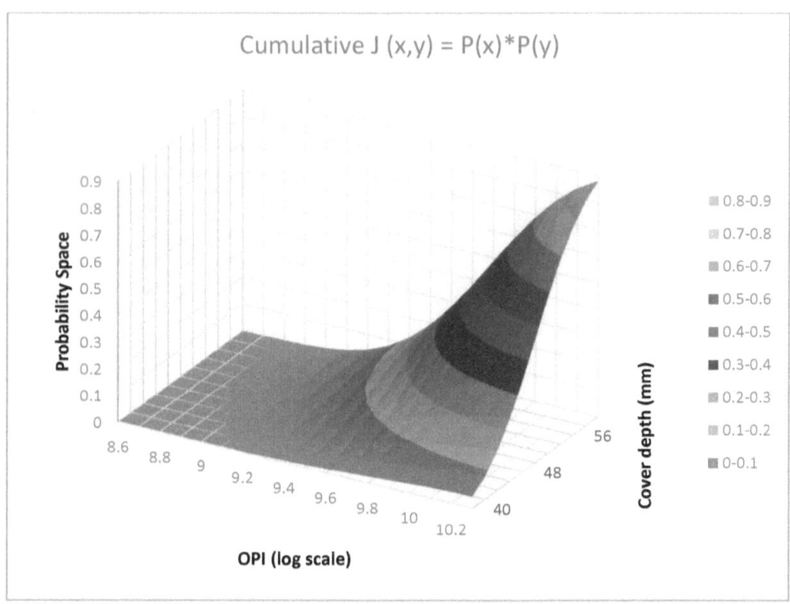

Figure 5-33 Cumulative Joint Probability Density Function (Project 3)

For Project 2, the joint probability of obtaining a set of values that satisfies both conditions equals 0.236, whereas for Project 3, the joint probability is much lower, equal to 0.066. Therefore, since the data for Project 2 represents test panels with a 23.6 % probability of achieving low OPI values and cover depth, the as-built quality should be verified by cores extracted from the actual structure since the probability will be expected to increase due to the trends identified in Project 5 between test panels and in-situ cores. A summary of the joint probability calculations for the limiting conditions in each project is included in *Table 5-12*.

Table 5-12 Joint Probability of OPI < 9.40 and Cover < 50 (Project 1 – 5)

	Project 1	Project 2	Project 3	Project 4		Project 5	
	TEST	TEST	TEST	TRIAL	TEST	TEST	CORES
Results	18	62	33	25	78	261	28
Determinations (x4)	72	248	132	100	312	1044	112
Mean OPI	9.43	9.36	9.79	10.57	9.94	9.67	9.26
Standard Deviation	0.43	0.22	0.35	0.23	0.29	0.29	0.31
Defectives (%)	44.44	61.29	12.12	3.77	5.66	17.61	71.43
Mean Cover Depth	50	50	50	50	50	50	50
Standard Deviation	6	6	6	6	6	6	6
P1 (OPI < 9.40)	0.4722	0.4722	0.1326	0.0000	0.0313	0.1759	0.6742
P2 (Cover < 50)	0.5	0.5	0.5	0.5	0.5	0.5	0.5
Joint Probability (P1 x P2)	0.2361	0.2361	0.0663	0.0000	0.0157	0.0880	0.3371

In Project 4, the joint probability of obtaining a set of values (DI and cover depth) satisfying both conditions (OPI < 9.40 and Cover < 50) only marginally increase when considering trial panels in relation to test panels due the relatively well performing specimens. However, in Project 5, the joint probability of obtaining a set of values (DI and cover depth) satisfying both conditions (OPI < 9.40 and Cover < 50) drastically increase by a factor of almost 4 when considering test panels in relation to cores.

6. Conclusions and recommendations

Assessing the variability of DI values such as the Oxygen Permeability Index (OPI) and Water Sorptivity Index (WSI) will assist in the differentiation of execution regimes that is evidently linked to curing and construction practices. Durability parameters are known to alter chiefly with age, composition, test method, execution and environment, hence durability data needs to be coupled with information on the above-mentioned variables.

The definition of test parameters for certain projects in terms of their single or joint probability distributions will assist in defining execution standards for different environmental regimes and concrete compositions. The collection of data using the proposed structure and the definition of parameters in terms of their mean and standard deviation will enable further use of the data in degradation models as the secondary purpose. However, the use of OPI values in degradation models, i.e. service-life estimates, is a primary purpose embodied in COTO (2018a; 2018b).

A reduction in the mean OPI value and increase in the standard deviation will increase the joint probability of obtaining low OPI and cover depth values. It is postulated that an increase in the mean WSI value and standard deviation will also increase the joint probability of obtaining high WSI and low cover depth values. Variations in the mean cover depth as well as the standard deviation will also affect the joint probability calculations, therefore it is pivotal that the cover depth values are assessed in line with DI values to translate the probability findings into a structure's possible risk of corrosion, given the exposure conditions, in a specified environment, for a concrete composition.

The ability to filter data will further enable information to be provided on the influence of the ageing effect and concrete composition on the durability parameters. Furthermore, collecting data from various test methods, execution and environmental regimes, will allow for verification of the test parameters (threshold limits) and deterioration rates linked to the exposure categories that form part of the degradation models. A database with results obtained from in-situ concrete and concrete cast under laboratory conditions also allows for the assessment of variability of performance of concrete with constant concrete composition.

Limiting values for durability parameters under early age testing regimes are the norm for performance-based specifications in South Africa. However, as new test data becomes available from different concrete compositions, test methods, execution and environmental regimes, these limiting values can be defined with greater confidence in line with the potential of concrete.

Substantially more data on Chloride Conductivity Index (CCI) is needed to determine the aging effect in the performance-based approach which is a very important aspect for a given concrete composition relating mainly to the kind of binder as well as the execution and exposure conditions. The aging factor to be applied on the diffusion coefficient is the most influential parameter when referring to concrete structures in marine environments. Therefore, various diffusion coefficients for chlorides obtained from different concrete compositions will allow for long term performance assessment of concrete structures.

However, since an outsized percentage of structures in the database are in inland environments, parameters such as Darcy's coefficient, and OPI are the most significant. Permeability coefficients for concrete obtained from different composition has the potential to allow for long term performance assessment of concrete structures with respect to carbonation-induced corrosion as well as defining project specific execution requirements.

Alexander, Bentur, & Mindess (2017) stated that Durability Indexes (DIs) primarily and empirically relate to Service Life Prediction (SLP) models. The DIs used are input parameters together with other variables such as cover and environmental class which determine the notional design life. Evidently, limiting DIs can and have been used in construction specifications to provide the necessary concrete quality for the required design life and environment. Two corrosion initiation models derived from measurements and correlations of short-term DI values, aggressiveness of environment and actual deterioration rates monitored up to 10 years.

Models allow for determining the expected life of a structure based on environmental conditions, cover thickness and concrete quality. The environmental classification is based on EN206-1 while concrete quality is represented by the appropriate DI parameter. SLMs can be used to determine the required value of the durability parameter based on pre-determined values for cover thickness, environment and expected design life. Alternatively, if the concrete quality is known, from the appropriate DI, a corrosion free life can be estimated for a given environment, but early-age to medium to long-term performance must be known.

6.1 Summary of observations and conclusions

Section 5.3 (Application of 'Deemed-to-satisfy' approach) highlighted the importance of assessing the mean value, CoV and relevant DI parameter to ascertain as-built quality. The range of average values for OPI from test panels varied from 9.36 (log) in Project 2 to 9.94 (log) in Project 4. From trial panels, the average OPI value for Project 3 was 10.57 (log). From cores, the average OPI value for Project 4 was 9.26 (log). OPI values along with the other material tests found within the database were assumed to be normally distributed which was supported by the data from all projects.

The CoV for Project 1 was smaller than that of Project 2, but due to the reduction of mean value for OPI and increase in mean value for WSI, more defectives were present in Project 1. Despite the increased CoV in Project 2, OPI and WSI were still within tolerances. In Project 4, the range of OPI and WSI values increased considering trial panels in relation to test panels as expected. Both mean OPI and WSI results displayed reductions in values from 10.57 (log) to 9.94 (log) and 5.97 mm/√hr to 5.54 mm/√hr, respectively. In the case of the latter, this proves that it is possible to achieve better quality concrete even despite the increased field variability encountered. The field variability can be quantified by the within CoV for both OPI and WSI which increased from 2.13 % to 2.87 % and 18.48 % to 28.25 %, respectively.

In Project 5, the mean OPI value shifted from 9.67 (log) to 9.26 (log), accompanied by an increase in CoV from 3.05 % to 3.39 % when considering test panels in relation to cores, which increased the amount of defectives from 17.99 % to 71.43 %. Furthermore, the mean WSI value also shifted from 7.16 mm/$\sqrt{hr}$ to 8.46 mm/$\sqrt{hr}$, accompanied by an increase in CoV from 23.36 % to 37.36 % when considering test panels in relation to cores, which further increased the amount of defectives from 13.79 % to 37.36 %.

This proves that in some circumstances, test panels do not sufficiently replicate the as-built quality compared to cores extracted from the actual structure. Despite compliance with the specification in terms of test panels (OPI value = 9.67 > 9.40), cores can contain OPI values exceeding the rejection limits in COTO (2018b). Evidently, identifying the relevant source of DI results will determine the correct proportion of defectives to be considered when applying the fixed payment adjustment factors to contract rates. The numerical summaries for the DI results are also of importance since this enables a partial or full recalculation of the original service life design assumptions to assess the residual service life of the structure (fib, 2006).

In order to predict the carbonation rate using OPI values, information from Module 2 (Concrete Composition), Module 3 (Execution) and Module 4 (Environment) need to be used. This information must also include the achieved cover depth (mm) in Module 6 (Test Results). Therefore, the system implemented will identify occurrences of defectives, poor OPI values on average (linked to the former) and low cover values.

The variability of durability properties for a given concrete is closely linked to the material, manufacturing and testing conditions, therefore the importance of defining parameters in terms of their mean and standard deviation is essential for a partial or full recalculation of the original service life design assumptions to assess the residual service life of the structure (fib, 2006). This contrasts with specifications for concrete durability such as in COTO (2018a; 2018b) which do not compute the variability in judgement plans. The partial factor method for carbonation-induced corrosion for uncracked concrete can be carried out by simple calculation without additional considerations regarding the probabilistic distributions regarding input parameters but is based on the full probabilistic design approach.

The full probabilistic method is based on Fick's law of diffusion which acts as the prevailing transport mechanism within concrete. However, the carbon dioxide diffusion coefficient is assumed to be a constant material property, even though for concrete it is a function of many variables (fib, 2006). The advantage of using this approach is that it leads to the most economical solutions, however, significantly larger expenses are encountered for the quantification of input parameters and the calculation itself.

The performance-based approach allows for better integration of concrete durability properties and considers variability in a rational way. Durability and corrosion degradation models assess the lifetime of concrete structures in a probabilistic way that considers the variability of all input parameters linked to the material, manufacturing and testing conditions. To enable the full probabilistic approach to be taken, an even bigger range of data is required.

This is needed firstly to verify that the law of probability density used is correct and secondly to quantify the input parameters in terms of their mean and standard deviation. Furthermore, with use of the database, certain parameters can be optimised to improve the performance from a technical and economic perspective that is in line with the rigorous approach for achieving concrete durability. Therefore, the creation of the physical database, through application of the designed data model (DM) is an essential tool for the advancement of an operational performance-based approach.

6.2 Summarized conclusions

The key question of the research was to measure or quantify the influences of site practices (material, manufacturing and testing conditions) such that inferences, and correlations could be made to actual in-situ performance.

The first specific key question involved subdividing the data into distinct groups or topics. which was covered in Chapter 3. This was completed by designing a conceptual data model (DM) such that DI results could be captured and structurally organised for further analysis based on the development of durability properties with change in material, manufacturing and testing conditions.

The second specific key question involved identifying the facts about each topic that need to be identified and stored, which was covered in Chapter 4. This was completed by designing a logical data model (DM) which added extra information to the conceptual data model (DM) elements. This process established the database tables or basic information required for the database which represents the structure of all data elements, set relationships between them, and provided a foundation to form the base for the physical database.

The last specific key question involved defining the relations between topics, which was covered in Chapter 5. This was completed by identifying relations between different input parameters which added extra information to the logical data model (DM) elements. In this penultimate chapter, the relations between topics were strengthened, which determined the extraction of information or output parameters from the physical database according to specification limits.

The physical creation of the database will deliver project specific numerical summaries of the key parameters that influence concrete durability which will be able to assist the evaluative process for conformity and decision-making process for action in new construction (acceptance, contractual penalties and remedial action). This database further will provide insight into the vast amount of durability test results from across the country that can be linked to successful material design characteristics and construction practices to inform on later improvements to achieve concrete durability. Secondly, data can be used for empirical and numerical detail-design procedures involved in rigorous approaches using DI values as input into SLM's for severe exposure conditions.

Therefore, the physical database will ultimately enable the long-term monitoring of our structures in a full-scale environment which will pave the way for further steps to be taken toward a fully probabilistic design approach.

If we are to shift toward the international paradigm of more durable concrete structures and rely on semi-invasive testing to control quality on site during and post-construction, an equal responsibility rests on concrete suppliers, contractors as well as consulting engineers as these three counterparts easily have the most pronounced impact on concrete durability at a hands-on level.

6.3 Recommendations

6.3.1 Practice

The advantages of integrating the Durability Index Database (DIDb) with an existing Bridge Management System (BMS) are threefold. In the long-term, development or improvement of performance relationships not modelled satisfactorily or that contain limitations can be undertaken. These would include the development or improvement of models that consider interdependent relationships between structure distress mechanisms such as the effect of cracking on accelerated corrosion. However, the immediate benefit of this system would involve the development of models that consider the effect of repair on future structure performance system. Examples of knowledge to be incorporated into the latter system would include the type of repair to be implemented, the conditions for applicability of the repair and what time the repair should be executed.

The integration of the DIDb with an existing BMS would allow for defects to be classified as durability or load-related and therefore further assessment can be undertaken on the effect of different types of structures (continuous, simply supported, integral or composite), construction (precast, cast-in-situ, balanced cantilever, cable stayed/suspension or arch) and material (reinforced concrete, pre-stressed concrete or steel) on repair strategies to increase durability and/or load-carrying capability.

6.3.2 Future research work

The physical database design should be executed following from the successful design and implementation of the conceptual and logical data models (DMs):

- The design of the base tables and integrity constraints using the available functionality of the target DBMS is the next step for the DIDb.

- The next step involves choosing the file organizations and indexes for the base tables. Typically, DBMSs provide several alternative file organizations for data, with the exception of PC DBMSs, which tend to have a fixed storage structure.

- The next step involves the design of the user views originally identified in the requirements analysis and collection stage of the database system development lifecycle.

- The integration of the DIDb with the SANRAL Bridge Management System (BMS).

Some of the above topics are part of current ongoing research.

References

(1999). Probabilistic methods for durability design : DuraCrete, probabilistic performance based durability design of concrete structures. [Gouda]: [CUR] .

Aït-Mokhtar, A., Belarbi, R., Benboudjema, F., Burlion, N., Capra, B., Carcassès, M., … Yanez-Godoy, H. (2013). Experimental investigation of the variability of concrete durability properties. *Cement and Concrete Research, 45*(1), 21–36. https://doi.org/10.1016/j.cemconres.2012.11.002

Alexander, M., Ballim, Y., & Kiliswa, M. (2013). Performance Testing and Specification of Concrete. In *Specification & Testing of Concrete Seminar*.

Alexander, M., Bentur, A., & Mindess, S. (2017). *Durability of Concrete: Design and Construction* (1st ed.). Boca Raton, FL 33487-2742: CRC Press.

Alexander, M. G., Ballim, Y., & Stanish, K. (2008). A framework for use of durability indexes in performance- based design and specifications for reinforced concrete structures. *Materials and Structures, 41*(2008), 921–936. https://doi.org/10.1617/s11527-007-9295-0

Anderson, D. A., Luhr, D. R., & Antle, C. E. (1990). National Cooperative Highway Research Program (NCHRP) Report 332: Framework for development of Performance-Related Specifications For Hot-Mix Asphaltic Concrete. *Transportation Research Board*. Washington, D.C.

Carcasses, M., Linger, L., Cussigh, F., Rougeau, P., Barberon, F., Thauvin, B., … Cubaynes, P. (2015). Setting Up of a Database Dedicated To Durability Indicators By the Civil Works French Association (AFGC) To Support the Implementation of Concrete Performance-Based Approach. In *Fib 2015* (pp. 1–11).

Cather, R. (1994, September). Curing: the True Story? *Magazine of Concrete Research V. 46 No. 168*, 157–161.

COLTO. (1998) Standard Specifications for Road and Bridge Works for State Authorities (Green Book). SAICE.

COTO. (2018a). Chapter 13: Structures. In *Standard Specifications for Road and Bridge Works* (Working Dr). Pretoria: The South African National Roads Agency SOC Limited.

COTO. (2018b). Chapter 20: Quality Assurance. In *Standard Specifications for Road and Bridge Works* (Working Dr). Pretoria: The South African National Roads Agency SOC Limited.

fib. (2006). *Bulletin 34: Model Code for Service Life Design*.

Foster, S. J., Stewart, M. G., Loo, M., Ahammed, M., & Sirivivatnanon, V. (2016). Calibration of Australian Standard AS3600 Concrete Structures: part I statistical analysis of material properties and model error. *Australian Journal of Structural Engineering, 17*(4), 242–253. https://doi.org/10.1080/13287982.2016.1246793

Gopinath, R., Alexander, M., & Beushausen, H. (2014). PREDICTING DEPTH OF CARBONATION OF CONCRETE - A PERFORMANCE-BASED APPROACH. In *RILEM International Symposium on Concrete Modelling* (pp. 12–14). Beijing.

Gouws, S., Alexander, M. G., & Maritz, G. (2001). Use of durability index tests for the assessment and control of concrete quality on site. *Concrete Beton, 98*, 5–16.

Gjørv, O. E. (2011). Durability of concrete structures. *Arabian Journal for Science and Engineering, 36*(2), 151–172. https://doi.org/10.1007/s13369-010-0033-5

Ibrahim, M., Shameem, M., Al-Mehthel, M., & Maslehuddin, M. (2013). Effect of curing methods on strength and durability of concrete under hot weather conditions. *Cement and Concrete Composites, 41*(2013), 60–69. https://doi.org/10.1016/j.cemconcomp.2013.04.008

Kessy, J. (2013). *Durability Specifications for Structural Concrete – an International Comparison.* University of Cape Town. Retrieved from http://uctscholar.uct.ac.za/PDF/102615_Kessy_J.pdf

Kessy, J. G., Alexander, M. G., & Beushausen, H. (2015). Concrete durability standards : International trends and the South African context. *Journal of the South African Institution of Civil Engineering, 57*(1), 47–58. https://doi.org/http://dx.doi.org/10.17159/2309-8775/2015/v57n1a5

Khan, M. U., Ahmad, S., & Al-Gahtani, H. J. (2017). Chloride-Induced Corrosion of Steel in Concrete: An Overview on Chloride Diffusion and Prediction of Corrosion Initiation Time. *International Journal of Corrosion, 2017.* https://doi.org/10.1155/2017/5819202

Kulkarni, V. R. (2009). Exposure classes for designing durable concrete. *Indian Concrete Journal, 83*(3), 23–43.

Li, K., Zhou, C., & Chen, Z. (2008). Chinese code for durability design of concrete structures : A state-of-art report. In *International Conference on Durability of Concrete Structures.* Hangzhou.

Linger, L., & Cussigh, F. (2018). PERFDUB: A New French Research Project on Performance-Based Approach for Justifying Concrete Structures Durability. In D. A. Hordijk & M. Luković (Eds.), *High Tech Concrete: Where Technology and Engineering Meet* (Vol. 1, pp. 2266–2274). Springer International Publishing. https://doi.org/10.1007/978-3-319-59471-2

Mekiso, F. A. (2013). Concrete Curing and its Practice in South Africa: A Literature Review. In *Test & Measurement NLA Conference* (pp. 1–11). Muldersdrift: National Laboratory Association.

Merretz, W. (2010). Achieving concrete cover in construction. *Concrete in Australia, 36*(1), 38–44.

Miyamoto, A., Katsushima, T., & Asano, H. (2013). Development of Practical Bridge Management System for Prestressed Concrete Bridges. *SMAR2013: Proceedings of the Second Conference on Smart Monitoring, Assessment and Rehabilitation of Civil Structures, SMAR 2013, 9th –11th September 2013, Istanbul, Turkey.*, (2000).

Miyamoto, A., & Nakamura, H. (2003). LCC Based Bridge Management System for Existing Concrete Bridges in Japan. In *ILCDES 2003: Integrated Lifetime Engineering of Buildings and Civil Infrastructures.* Rotterdam (Netherlands).

Moyana, P. (2015). A Database of Durability Test Results. University of Cape Town.

Nell, A. J., Newmark, A., & Nordengen, P. A. (2008). A Bridge Management System for the Western Cape Provincial Government, South Africa: Implementation and Utilisation.

Nganga, G. W. (2011). *Practical Implementation of the Durability Index-Based Performance Approach.* University of Cape Town.

Nganga, G. W., Alexander, M. G., & Beushausen, H. B. (2017). Practical Implementation of Durability Index Performance-based Specifications: Current Experience.

Nganga, G. W., Alexander, M. G., & Beushausen, H. D. (2014). QUALITY CONTROL OF REINFORCED CONCRETE STRUCTURES. In *XIII Conference on Durability of Building Materials and Components* (pp. 1006–1014).

Obla, K. H. (2014). *Improving Concrete Quality* (1st ed.). Boca Raton, FL 33487-2742: CRC Press. Retrieved from http://dundee.eblib.com/patron/FullRecord.aspx?p=1418402

Qi, G., Niu, D., Yuan, C., & Duan, F. (2012). Research on the Variation of pH Value of the Ordinary Concrete and Fly Ash Concrete in the Carbonation Process. *Advanced Materials Research*, *374–377*, 1934–1937. https://doi.org/10.4028/www.scientific.net/AMR.374-377.1934

Ronny, R. (2011). *Critique of Durability Specifications for Concrete Bridges on National Roads in South Africa*.

SANRAL. (2009). Table 6000 / 1 : Concrete Durability Specification Targets (Civil Enginering Structures only) for Carbonation-Induced Corrosion (from Atmospheric & Industrial) and Chloride-Induced Corrosion (from Groundwater , Seawater & Sea spray). *Concrete*.

SANRAL. (2014). Chapter 3: Materials Testing. In *South African Pavement Engineering Manual*.

SANRAL. (2016). *Annual Report*.

SANRAL. (2017). SANRAL - ITIS 5.1. Retrieved September 1, 2017, from https://itis.nra.co.za/Portal/

SANRAL Works Contract Proforma. (2019). Version 2019.1 – 19 July 2019.

Schueremans, L. & Van Gemert, D., 1997. *Service Life Prediction Model for Reinforced Concrete Constructions Treated with Water-repellent Compounds,* Brussels: Ministerie van de Vlaamse Gemeenschap.

Stanish, K., Alexander, M. G., & Ballim, Y. (2006). Assessing the repeatability and reproducibility values of South African. *Journal of the South African Institution of Civil Engineering*, *48*(2), 10–17.i

Surana, S., Pillai, R. G., & Santhanam, M. (2017). Performance evaluation of curing compounds using durability parameters. *Construction and Building Materials*, *148*(2017), 538–547. https://doi.org/10.1016/j.conbuildmat.2017.05.055

UCT. (2018a). Durability Index Spreadsheet Template. Cape Town.

UCT. (2018b). Durability Index Testing Procedure Manual (Vol. Version 4.). Cape Town.

Van Wyk, E., Van Tonder, G., & Vermeulen, D. (2012). Characteristics of local groundwater recharge cycles in South African semi-arid hard rock terrains: Rainfall–groundwater interaction

Visser, J. H. M., & Han, N. (2003). *TNO Report 2002-CI-R2012 Database for durability properties of Concrete Design & manual*. Delft. https://doi.org/006.10176/01.03.01

Appendix A: DIDb Conceptual Framework

Key

1. <u>Concrete Durability Module (CDM):</u> Module envisaged for capturing DI test results within the South African Road Design Software (SARDS)
2. <u>Durability Index (DI) Approach:</u> Quantifiable engineering parameter characterising concrete cover quality sensitive to material, manufacturing and testing conditions
3. <u>Laboratory conditions:</u> Experimental conditions representing the material potential of concrete
4. <u>Integrated Transportation Information System (ITIS):</u> Comprehensive database tool developed by SANRAL involving various functions such as definition (schema), creation (tables), querying (reports and views), update (user interface) and administration (modification)
5. <u>Field conditions:</u> Observational conditions representing the as-built quality of concrete
6. <u>Recommended values:</u> Target design values for concrete mix design (Table 6000/1: Concrete Durability Specification Targets)
7. <u>Numerical summary:</u> Range (maximum – minimum), target mean values, standard deviation and coefficient of variation (CoV)
8. <u>Between CoV:</u> Measure of variance of one operator conducting a test on a material and repeating the test (repeatability and reliability)
9. <u>Within CoV:</u> Measure of variance of one material in same environment due to different manufacturing conditions – illustrated by distribution
10. <u>Specification Limit:</u> Limit value outside which not more than a certain specified percentage (Φ) of the population of values representing an acceptable property is allowed to lie - Single lower limit Ls (OPI), or single upper limit L's (WSI)
11. <u>Defectives:</u> Expressed as a percentage of the population of values

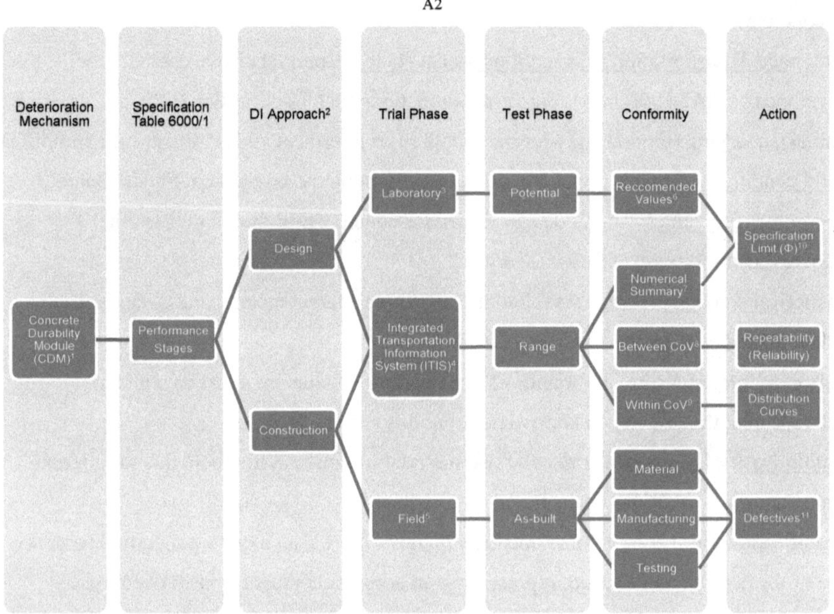

Deterioration Mechanism
Specification Table 6000/1
DI Approach[2]
Trial Phase
Test Phase
Conformity
Action
Concrete Durability Module (CDM)[1]
Performance Stages
Design
Construction
Laboratory[3]
Integrated Transportation Information System (ITIS)[4]
Field[5]
Potential
Range
As-built
Reccomended Values[6]
Numerical Summary[7]
Between CoV[8]
Within CoV[9]
Material
Manufacturing
Testing
Specification Limit (Φ)[10]
Repeatability (Reliability)
Distribution Curves
Defectives[11]

Appendix B: Lucidchart Entity Relationship Diagram (ERD)

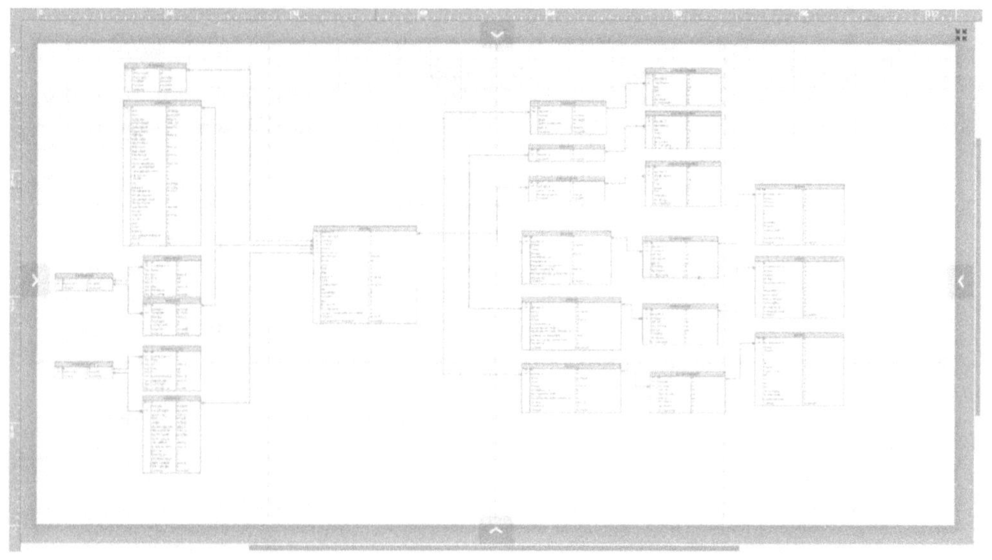

Type 1: Composition / Reference and Specimen

One (and only one): Composition / Reference — One or many: Specimen

Type 2: Regimes, Details and Execution / Environment, Specimen

One (and only one): Exposure / Curing Regime — One or many: Exposure / Curing details

One or many: Execution / Environment

One or many: Specimens

Type 3: Cover, Mass and Compressive Strength

One or many: Specimen — One (and only one): Detail — One or many: Measurements

Type 4: OPI, CCI and WSI

Many: Specimen — One (and only one): Detail — Many: Measurements: One (and only 1) = One (and only 1) Data

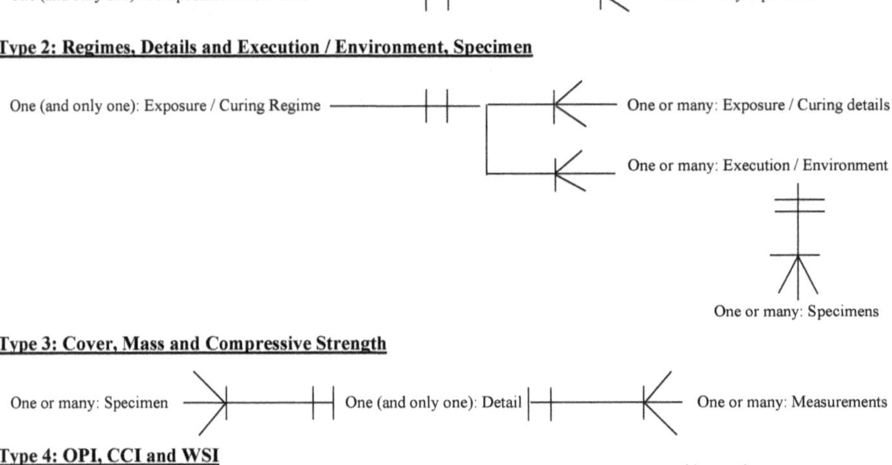

Appendix C: Common Cements Table (SANS 50197-1)

Main types	Notation of products (types of common cement)		Composition, percentage by mass[a]										
			Clinker	Blast-furnace slag	Silica fume	Pozzolana Natural	Pozzolana Natural calcined	Fly ash Siliceous	Fly ash Calcareous	Burnt shale	Limestone	Limestone	Minor additional constituents
			K	S	D[b]	P	Q	V	W	T	L	LL	
CEM I	Portland cement	CEM I	95 - 100	-	-	-	-	-	-	-	-	-	0 - 5
CEM II	Portland-slag cement	CEM II A-S	80 - 94	6 - 20	-	-	-	-	-	-	-	-	0 - 5
		CEM II B-S	65 - 79	21 - 35	-	-	-	-	-	-	-	-	0 - 5
	Portland-silica fume cement	CEM II A-D	90 - 94	-	6 - 10	-	-	-	-	-	-	-	0 - 5
	Portland-pozzolana cement	CEM II A-P	80 - 94	-	-	6 - 20	-	-	-	-	-	-	0 - 5
		CEM II B-P	65 - 79	-	-	21 - 35	-	-	-	-	-	-	0 - 5
		CEM II A-Q	80 - 94	-	-	-	6 - 20	-	-	-	-	-	0 - 5
		CEM II B-Q	65 - 79	-	-	-	21 - 35	-	-	-	-	-	0 - 5
	Portland-fly ash cement	CEM II A-V	80 - 94	-	-	-	-	6 - 20	-	-	-	-	0 - 5
		CEM II B-V	65 - 79	-	-	-	-	21 - 35	-	-	-	-	0 - 5
		CEM II A-W	80 - 94	-	-	-	-	-	6 - 20	-	-	-	0 - 5
		CEM II B-W	65 - 79	-	-	-	-	-	21 - 35	-	-	-	0 - 5
	Portland-burnt shale cement	CEM II A-T	80 - 94	-	-	-	-	-	-	6 - 20	-	-	0 - 5
		CEM II B-T	65 - 79	-	-	-	-	-	-	21 - 35	-	-	0 - 5
	Portland-limestone cement	CEM II A-L	80 - 94	-	-	-	-	-	-	-	6 - 20	-	0 - 5
		CEM II B-L	65 - 79	-	-	-	-	-	-	-	21 - 35	-	0 - 5
		CEM II A-LL	80 - 94	-	-	-	-	-	-	-	-	6 - 20	0 - 5
		CEM II B-LL	65 - 79	-	-	-	-	-	-	-	-	21 - 35	0 - 5
	Portland-composite cement[c]	CEM II A-M	80 - 88	←				12 - 20				>	0 - 5
		CEM II B-M	65 - 79	←				21 - 35				→	0 - 5
CEM III	Blast furnace cement	CEM III A	35 - 64	36 - 65	-	-	-	-	-	-	-	-	0 - 5
		CEM III B	20 - 34	66 - 80	-	-	-	-	-	-	-	-	0 - 5
		CEM III C	5 - 19	81 - 95	-	-	-	-	-	-	-	-	0 - 5
CEM IV	Pozzolanic cement[c]	CEM IV A	65 - 89	-	←		11 - 35	>	-				0 - 5
		CEM IV B	45 - 64	-	←		36 - 55	>					0 - 5
CEM V	Composite cement[c]	CEM V A	40 - 64	18 - 30	-	←	18 - 30	>	-	-	-	-	0 - 5
		CEM V B	20 - 38	31 - 49	-	←	31 - 49	>	-	-	-	-	0 - 5

Notes

(a) The values in the table refer to the sum of the main and minor additional constituents.

(b) The proportion of silica fume is limited to 10%.

(c) In portland-composite cements CEM II A-M and CEM II B-M, in pozzolanic cements CEM IV A and CEM IV B, and in composite cements CEM V A and CEM V B, the main constituents other than clinker shall be declared by designation of the cement.

Appendix D: Verification of Durability Specification

Table A-1: Nominal Durability Index and cover values for 100-year service life in typical carbonating environme.
(Table A13.4.7-3) - COTO (2018a) & SANRAL (2009)

Environmental class	Cover (mm), as specified	OPI (log scale)	Table 6000/1	Table 6000/1
For 100 year service life			Recommended	Minimum
XC1a, and	40	9.15	9.20	9.00
XC1b	50	9.00	9.00	9.00
	60	9.00	n/a	n/a
XC2	40	9.40	9.40	9.00
	50	9.10	9.10	9.00
	60	9.00	9.00	9.00
XC3	40	9.65	9.40	9.00
	50	9.35	9.10	9.00
	60	9.05	9.00	9.00
XC4	40	9.85	9.60	9.20
	50	9.55	9.30	9.00
	60	9.30	9.10	9.00
	(70)		9.00	9.00

Table A-2: Nominal Durability Index and cover values for 100-year service life in typical chloride environment
(Table A13.4.7-3) - COTO (2018a) & SANRAL (2009)

Environmental class	Cover (mm), as specified	Chloride Conductivity (mS/cm)			
		Typical Cementitious Binder System			
		Fly ash (30 %)	Blastfurnace slag (50 %)	Corex slag (50 %)	Silica fume (10 %)
For 100 year service life					
XS1	40	1.20 (1.50)	1.30 (1.60)	1.60^2 (2.10)	n/a^1 (0.40)
	50	1.85^2 (2.10)	1.95^2 (2.20)	2.20^2 (2.80)	0.40 (0.50)
	60	2.15^2 (2.60)	2.35^2 (2.70)	2.75^2 (3.40)	0.65 (0.65)
XS2a	(40)	- (1.00)	- (1.10)	- (1.40)	- (0.30)
	50	0.85 (1.40)	1.00 (1.60)	1.20 (2.00)	n/a^1 (0.40)
	60	1.25 (1.80)	1.45^2 (2.10)	1.70^2 (2.50)	n/a^1 (0.50)
XS2b	60	1.10 (1.45)	1.30 (1.70)	1.55^2 (2.00)	n/a^1 (0.40)
XS3a	(40)	- (0.65)	- (0.85)	- (1.00)	- (0.25)
	50	0.65 (1.10)	0.80 (1.35)	0.95 (1.45)	n/a^1 (0.35)
	60	0.95 (1.45)	1.10 (1.70)	1.40 (2.00)	n/a^1 (0.40)
XS3b	60	0.85 (1.10)	1.00 (1.30)	1.30 (1.55)	n/a^1 (0.30)

Notes: *1   n/a means cementitious binder system is not suitable for the indicated purpose*
2   Maximum water: cementitious binder ratio for all binder systems shall be maximum 0.550

Table A-3: Carbonation Depths for upper OPI limit (100-year design service life)

Environmental Class	Environment Category	OPI (Upper)	PC / BS / CS	Δ	>Cover	FA / SF	Δ	>Cover
XC1a - 40mm	20 – Coastal	9.15	34.6mm	5.6	✗	47.7mm	-7.7	✓
50mm	(Av. R.H = 80%)	9.00	38.4mm	11.6	✗	53.0mm	-3.0	✓
60mm		9.00	38.4mm	21.6	✗	53.0mm	7.0	✗
XC1b - 40mm	30 – Partly wet	9.15	17.3mm	22.7	✗	23.9mm	16.1	✗
50mm	(Av. R.H = 90%)	9.00	19.2mm	30.8	✗	26.5mm	23.5	✗
60mm		9.00	19.2mm	30.8	✗	26.5mm	33.5	✗
XC2 - 40mm	30 – Partly wet	9.40	14.1mm	25.9	✗	19.4mm	20.6	✗
50mm	(Av. R.H = 90%)	9.10	17.9mm	32.1	✗	24.7mm	32.1	✗
60mm		9.00	19.2mm	40.8	✗	26.5mm	40.8	✗
XC3 - 40mm	10 – Dry inland	9.65	29.8mm	10.2	✗	41.2mm	-1.2	✓
50mm	(Av. R.H = 60%)	9.35	40.4mm	9.6	✗	55.7mm	-5.7	✓
60mm		9.05	50.9mm	9.1	✗	70.3mm	-10.3	✓
XC4 - 40mm	10 – Dry inland	9.85	22.8mm	17.2	✗	31.5mm	8.5	✗
50mm	(Av. R.H = 60%)	9.55	33.3mm	16.7	✗	46.0mm	4.0	✗
60mm		9.30	42.1mm	17.9	✗	58.1mm	1.9	✗

Table A-4: Carbonation Depths for lower OPI limit (100-year design service life)

Environmental Class	Environment Category	OPI (Lower)	PC / BS / CS	Δ	>Cover	FA / SF	>Cover	Δ
XC1a - 40mm	20 – Coastal	8.90	41.0mm	-1.0	✓	56.5mm	✓	-16.5
50mm	(Av. R.H = 80%)	8.75	44.8mm	5.2	✗	61.8mm	✓	-11.8
60mm		8.75	44.8mm	15.2	✗	61.8mm	✓	-1.8
XC1b - 40mm	30 – Partly wet	8.90	20.5mm	19.5	✗	28.3mm	✗	11.7
50mm	(Av. R.H = 90%)	8.75	22.4mm	27.6	✗	30.9mm	✗	19.1
60mm		8.75	22.4mm	37.6	✗	30.9mm	✗	29.1
XC2 - 40mm	30 – Partly wet	9.15	17.3mm	22.7	✗	23.9mm	✗	16.1
50mm	(Av. R.H = 90%)	8.85	21.1mm	28.9	✗	29.2mm	✗	20.8
60mm		8.75	22.4mm	37.6	✗	30.9mm	✗	29.1
XC3 - 40mm	10 – Dry inland	9.40	38.6mm	1.4	✗	53.3mm	✓	-13.3
50mm	(Av. R.H = 60%)	9.10	49.1mm	0.9	✗	67.8mm	✓	-17.8
60mm		8.80	59.7mm	0.3	✗	82.4mm	✓	-22.4
XC4 - 40mm	10 – Dry inland	9.60	31.6mm	8.4	✗	43.6mm	✓	-3.6
50mm	(Av. R.H = 60%)	9.30	42.1mm	7.9	✗	58.1mm	✓	-8.1
60mm		9.05	50.9mm	9.1	✗	70.3mm	✓	-10.3

Table A-4: Threshold Chloride Content (0.4 %) Depth for lower CCI limit (100-year design service life) – Fly Ash & Blast furnace Slag

Environmental Class	Exposure Category	CCI (Lower)	FA (30%)	Δ	>Cover	CCI (Lower)	BS (50%)	Δ	>Cover
XS1 - 40mm	30 - Severe	1.20	35.0mm	5.0	✗	1.30	35.0mm	5.0	✗
50mm		1.85	44.0mm	6.0	✗	1.95	44.0mm	6.0	✗
60mm		2.15	50.0mm	10.0	✗	2.35	51.0mm	11.0	✗
XS2a - 50mm	20 – Very severe	0.85	43.5mm	6.5	✗	1.00	42.5mm	7.5	✗
60mm		1.25	53.0mm	7.0	✗	1.45	52.5mm	7.5	✗
XS2b - 60mm		1.10	49.0mm	11.0	✗	1.30	49.0mm	11.0	✗
XS3a - 50mm	10 - Extreme	0.65	42.0mm	8.0	✗	0.80	42.5mm	7.5	✗
60mm		0.95	51.5mm	8.5	✗	1.10	51.5mm	8.5	✗
XS3b - 60mm		0.85	48.0mm	12.0	✗	1.00	48.0mm	12.0	✗

Table A-5: Threshold Chloride Content (0.4 %) Depth for lower CCI limit (100-year design service life) – Corex Slag & Silica Fume

Environmental Class	Exposure Category	CCI (Lower)	CS (50%)	Δ	>Cover	CCI (Lower)	SF (10%)	Δ	>Cover
XS1 - 40mm	30 - Severe	1.60	34.0mm	6.0	✗	-	-	-	-
50mm		2.20	41.0mm	9.0	✗	0.40	39.0mm	11.0	✗
60mm		2.75	48.0mm	12.0	✗	0.65	44.0mm	16.0	✗
XS2a - 50mm	20 – Very severe	1.20	43.0mm	7.0	✗	-	-	-	-
60mm		1.70	52.0mm	8.0	✗	-	-	-	-
XS2b - 60mm		1.55	48.5mm	11.5	✗	-	-	-	-
XS3a - 50mm	10 - Extreme	0.95	42.5mm	7.5	✗	-	-	-	-
60mm		1.40	52.0mm	8.0	✗	-	-	-	-
XS3b - 60mm		1.30	50.0mm	10.0	✗	-	-	-	-

Table A-6: Threshold Chloride Content (0.4 %) Depth for upper CCI limit (100-year design service life) – Fly Ash & Blast furnace Slag

Environmental Class	Exposure Category	CCI (Upper)	FA (30%)	Δ	>Cover	CCI (Upper)	BS (50%)	Δ	>Cover
XS1 - 40mm	30 - Severe	1.40	38.0mm	2.0	✗	1.50	37.0mm	3.0	✗
50mm		2.05	49.0mm	1.0	✗	2.15	48.0mm	2.0	✗
60mm		2.35	55.0mm	5.0	✗	2.55	55.0mm	5.0	✗
XS2a - 50mm	20 – Very severe	1.05	48.5mm	1.5	✗	1.20	47.0mm	3.0	✗
60mm		1.45	58.5mm	1.5	✗	1.65	57.0mm	3.0	✗
XS2b - 60mm		1.30	55.0mm	5.0	✗	1.50	54.0mm	6.0	✗
XS3a - 50mm	10 - Extreme	0.85	48.0mm	2.0	✗	1.00	48.5mm	1.5	✗
60mm		1.15	58.0mm	2.0	✗	1.30	57.5mm	2.5	✗
XS3b - 60mm		1.05	55.0mm	5.0	✗	1.20	54.0mm	6.0	✗

Table A-5: Threshold Chloride Content (0.4 %) Depth for upper CCI limit (100-year design service life) – Corex Slag & Silica Fume

Environmental Class	Exposure Category	CCI (Upper)	CS (50%)	Δ	>Cover	CCI (Upper)	SF (10%)	Δ	>Cover
XS1 - 40mm	30 - Severe	1.80	37.0mm	3.0	✗	-	-	-	-
50mm		2.40	44.0mm	6.0	✗	0.60	55.0mm	-5.0	✓
60mm		2.95	51.0mm	9.0	✗	0.85	64.0mm	-14.0	✓
XS2a - 50mm	20 – Very severe	1.40	46.0mm	4.0	✗	-	-	-	-
60mm		1.90	55.0mm	5.0	✗	-	-	-	-
XS2b - 60mm		1.75	53.0mm	7.0	✗	-	-	-	-
XS3a - 50mm	10 - Extreme	1.15	47.0mm	3.0	✗	-	-	-	-
60mm		1.60	57.0mm	3.0	✗	-	-	-	-
XS3b - 60mm		1.50	55.0mm	5.0	✗	-	-	-	-

Appendix E: Ethics Signature Form

Application for Approval of Ethics in Research (EIR) Projects
Faculty of Engineering and the Built Environment, University of Cape Town

APPLICATION FORM

Please Note:

Any person planning to undertake research in the Faculty of Engineering and the Built Environment (EBE) at the University of Cape Town is required to complete this form *before* collecting or analysing data. The objective of submitting this application *prior* to embarking on research is to ensure that the highest ethical standards in research, conducted under the auspices of the EBE Faculty, are met. Please ensure that you have read, and understood the EBE Ethics in Research Handbook (available from the UCT EBE, Research Ethics website) prior to completing this application form: http://www.ebe.uct.ac.za/ebe/research/ethics1

APPLICANT'S DETAILS		
Name of principal researcher, student or external applicant	Daniel Govender	
Department	Civil Infrastructure Management & Maintenance	
Preferred email address of applicant	GVNDAN004@myuct.ac.za	
If Student	Your Degree: e.g., MSc, PhD, etc.	MSc (Eng)
	Credit Value of Research: e.g., 60/120/180/360 etc.	120
	Name of Supervisor (if supervised):	Emeritus Professor Mark Alexander
If this is a research contract, indicate the source of funding/sponsorship	N/A	
Project Title	Application of the Durability Index (DI) approach in industry	

I hereby undertake to carry out my research in such a way that:

- there is no apparent legal objection to the nature or the method of research; and
- the research will not compromise staff or students or the other responsibilities of the University;
- the stated objective will be achieved, and the findings will have a high degree of validity;
- limitations and alternative interpretations will be considered;
- the findings could be subject to peer review and publicly available; and
- I will comply with the conventions of copyright and avoid any practice that would constitute plagiarism.

SIGNED BY	Full name	Signature	Date
Principal Researcher/ Student/External applicant	Daniel Govender	*Signature Removed*	12 Sep 2018

APPLICATION APPROVED BY	Full name	Signature	Date
Supervisor (where applicable)	Emeritus Professor Mark Alexander	*Signature Removed*	12 Sep 2018
HOD (or delegated nominee) Final authority for all applicants who have answered NO to all questions in Section1; and for all Undergraduate research (Including Honours).	Professor Pilate Moyo	*Signature Removed*	12 Sep 2018
Chair : Faculty EIR Committee For applicants other than undergraduate students who have answered YES to any of the above questions.	Dyllon Randall *Signature Removed*		17 Oct 2018

www.ingramcontent.com/pod-product-compliance
Lightning Source LLC
LaVergne TN
LVHW042149190726
843493LV00006B/1577